A Naturalist's

San Juan RIVER GUIDE

Waterproof Edition

The Goosenecks — Michael Collier and Chris Condit

TEXT: Stewart Aitchison
GEOLOGY: Peter Winn
ARCHAEOLOGY: Don Keller
ILLUSTRATIONS: Barbara Kemp

PRUETT PUBLISHING COMPANY
Boulder, Colorado

FRONT COVER PHOTO: The Goosenecks—Michael Collier and Chris Condit
BACK COVER PHOTO: Datura—John Running, Wild & Scenic

First Edition
1 2 3 4 5 6 7 8 9

Printed in the United States of America

Library of Congress Cataloging in Publication Data
Aitchison, Stewart W.
A naturalist's San Juan River guide.

Bibliography: p.
Includes index.
1. Natural history—San Juan River (Colo.-Utah)—Guide-books. 2. San Juan River (Colo.-Utah)—Description and travel—Guide-books. I. Title.
QH104.5.S25A37 508.792'59 82-3718
ISBN: 0-87108-653-0 (waterproof) AACR2

Grateful acknowledgement is made for permission to reprint:

Excerpt from *The Ghosts of the Old San Juan* by Katie Lee. Reprinted by permission of Katie Lee and Folkways Records.

Excerpt from *The Sound of Mountain Water* by Wallace Stegner. Reprinted by permission of Doubleday and Company, Inc.

Excerpt from "Short Cut to the San Juan" by Charles Redd from *1949 Brand Book*. Reprinted by permission of the Denver Posse of the Westerners.

Excerpt from *Wind in the Rock* by Ann Zwinger. Reprinted by permission of Harper & Row, Publishers, Inc.

Excerpt from *Run, River, Run* by Ann Zwinger. Reprinted by permission of Harper & Row, Publishers, Inc.

Datura John Running, Wild & Scenic

May Our Children Have
Wild Rivers To Run

There's a legend they tell of treasures galore
Along the old San Juan
Where there's long lost plunder in deep hidden store
Along the old San Juan
Well, them that searched have come and gone
There ain't nothin' left but the old San Juan
And the ghosts of a wild old river.

Katie Lee, 1964

To start a trip at Mexican Hat, Utah, is to start off into empty space from the end of the world. The space that surrounds Mexican Hat is filled only with what the natives describe as "a lot of rocks, a lot of sand, more rocks, more sand, and wind enough to blow it away."

Wallace Stegner, 1946

Anasazi Petroglyphs John Running, Wild & Scenic

CONTENTS

LOCATION MAP

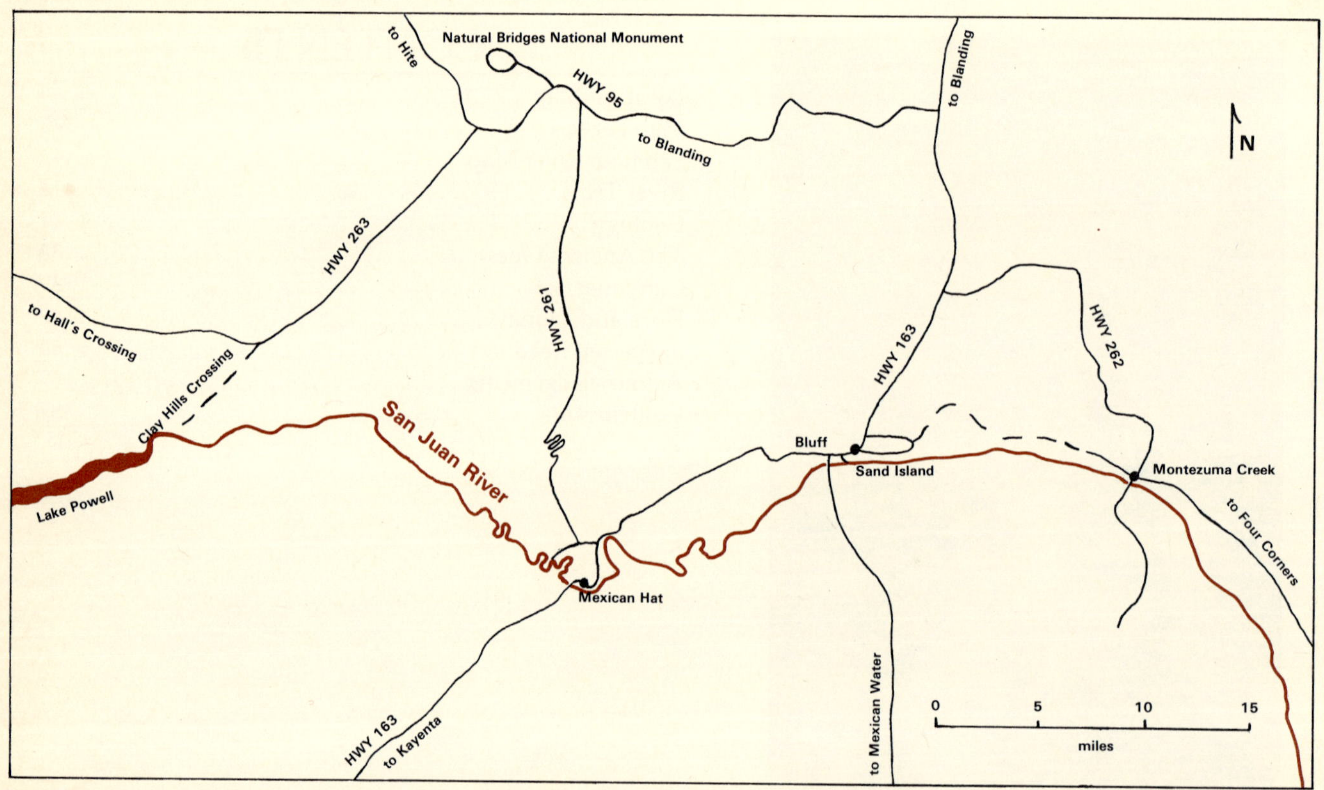

MAP LEGEND

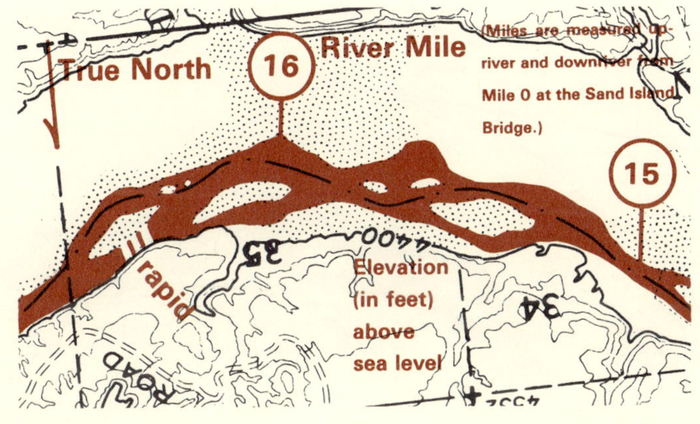

CAUTION: It is assumed that the user of this river guide is well-versed in whitewater navigation. Although the rapids found on the San Juan are not very difficult relative to other southwestern rivers, the steep gradient (and resultant fast speed) of the river, hidden sandbars, snags, and sudden storms or changes in runoff can create dangerous situations.
Be prepared! Be experienced! Stay alive!

The river map is based on U.S. Geological Survey 1953–63 Fifteen Minute Series topographic maps.

If this river guide gets wet, separate pages and let dry before storing.

SAN JUAN RIVER MAP

From J.N. Macomb, 1876

San Juan Mountains

Melting snows high in the majestic peaks of the San Juan Mountains of southwestern Colorado give rise to a major tributary of the Colorado River, the San Juan. The San Juan River flows south into New Mexico then briefly back into Colorado before turning westward into Utah. Once in the slickrock country of southeastern Utah,

19

16

15

N

18

17

River Access

River Access

the river becomes entrenched in a series of deep, awesome canyons as it meanders to its death in the great eastern arm of Lake Powell.

The spectacular Goosenecks of the San Juan are a familiar sight to southwestern travelers; and this section of the Colorado Plateau is particularly rich in prehistoric and historic legend and fact.

From Bluff, Utah, to Grand Gulch, the San Juan is still a wild, beautiful river. So dip your paddle into the muddy chocolate or sometimes clear amber waters; let your spirit soar over the canyon; and drink in the elixir of desert sun and glowing rock.

In the course of our march we observed many ruins of houses and found quantities of fragments of pottery scattered over the ground, indicating that the valley was once occupied by a race probably of the same origin and character as the Pueblo Indians extant in New Mexico. The fate of these former occupants of that dreary region is involved in mystery.

J. N. Macomb, 1876

Anasazi Ruin W.H. Jackson

From J.N. Macomb, 1876

Monument Valley

and scattered over the interval are many castle-like buttes and slender towers . . . their forms wonderful imitations of the structures of human art. Illuminated by the setting sun, the outlines of these singular objects came out sharp and distinct . . . we could hardly resist the conviction that we beheld the walls and towers of some Cyclopean city hitherto undiscovered in this far-off region.

J. S. Newberry, 1876

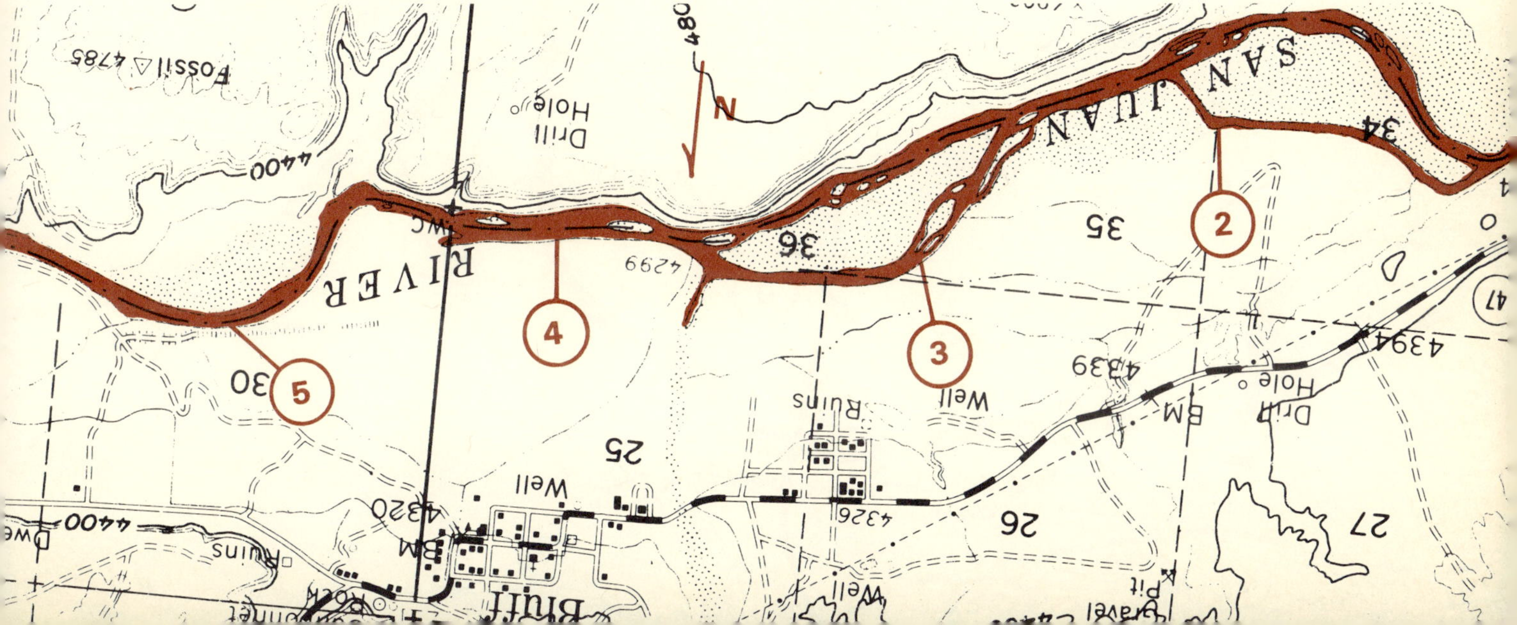

Every conceivable kind of watercraft has been down the San Juan River. From inner tubes to driftwood rafts, kayaks to sportyaks, rowboats to motorboats, adventurous boatmen have tested their skill against the rapids, sand waves, and the river gods.

Canoeists

Stewart Aitchison

Sand Island Campground & River Access

Sand Island Bridge

Anasazi Steps

AIRSTRIP

Stewart Aitchison

Mormon Wagon

Aside from the Hole-in-the-Rock, itself, this (San Juan Hill) was the steepest crossing on the journey. Here again seven span of horses were used, so that when some of the horses were on their knees, fighting to get up to find a foothold, the still erect horses could plunge upward. . . .My father was a strong man and reluctant to display emotion; but whenever in later years the full pathos of San Juan Hill was recalled . . .he wept.

Charles Redd, 1950

4

5

6

7

Anasazi Petroglyphs

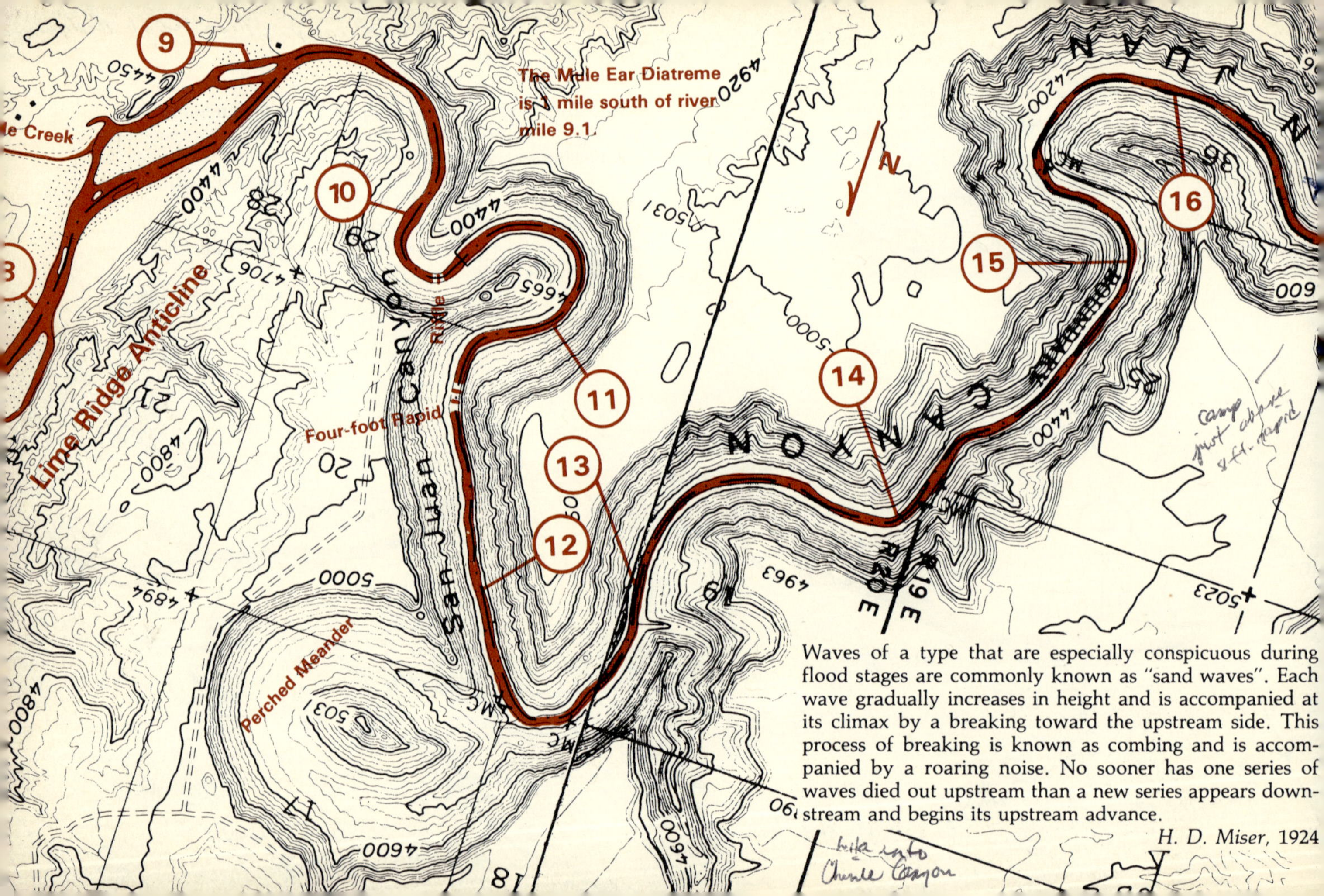

Waves of a type that are especially conspicuous during flood stages are commonly known as "sand waves". Each wave gradually increases in height and is accompanied at its climax by a breaking toward the upstream side. This process of breaking is known as combing and is accompanied by a roaring noise. No sooner has one series of waves died out upstream than a new series appears downstream and begins its upstream advance.

H. D. Miser, 1924

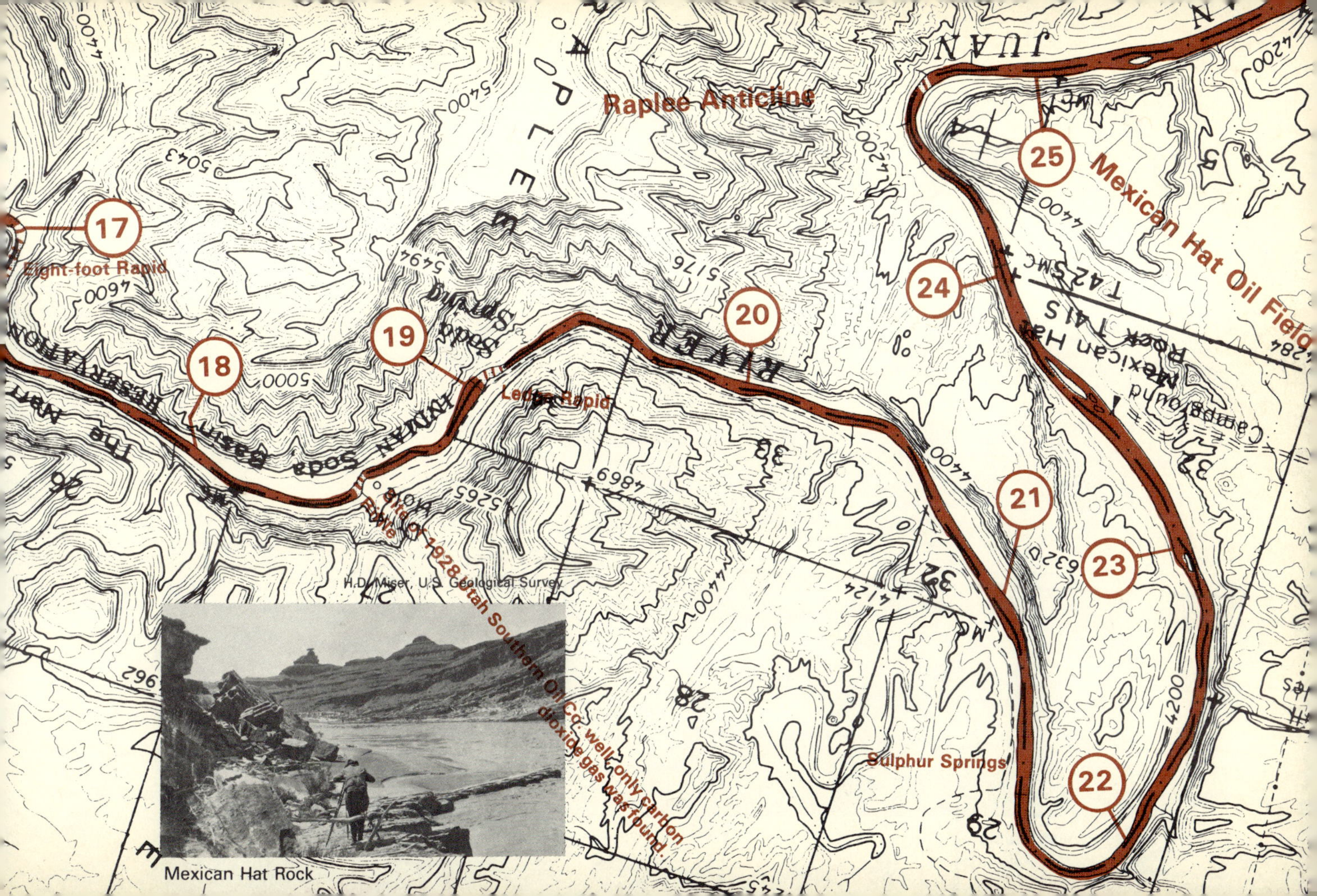

Mexican Hat Rock

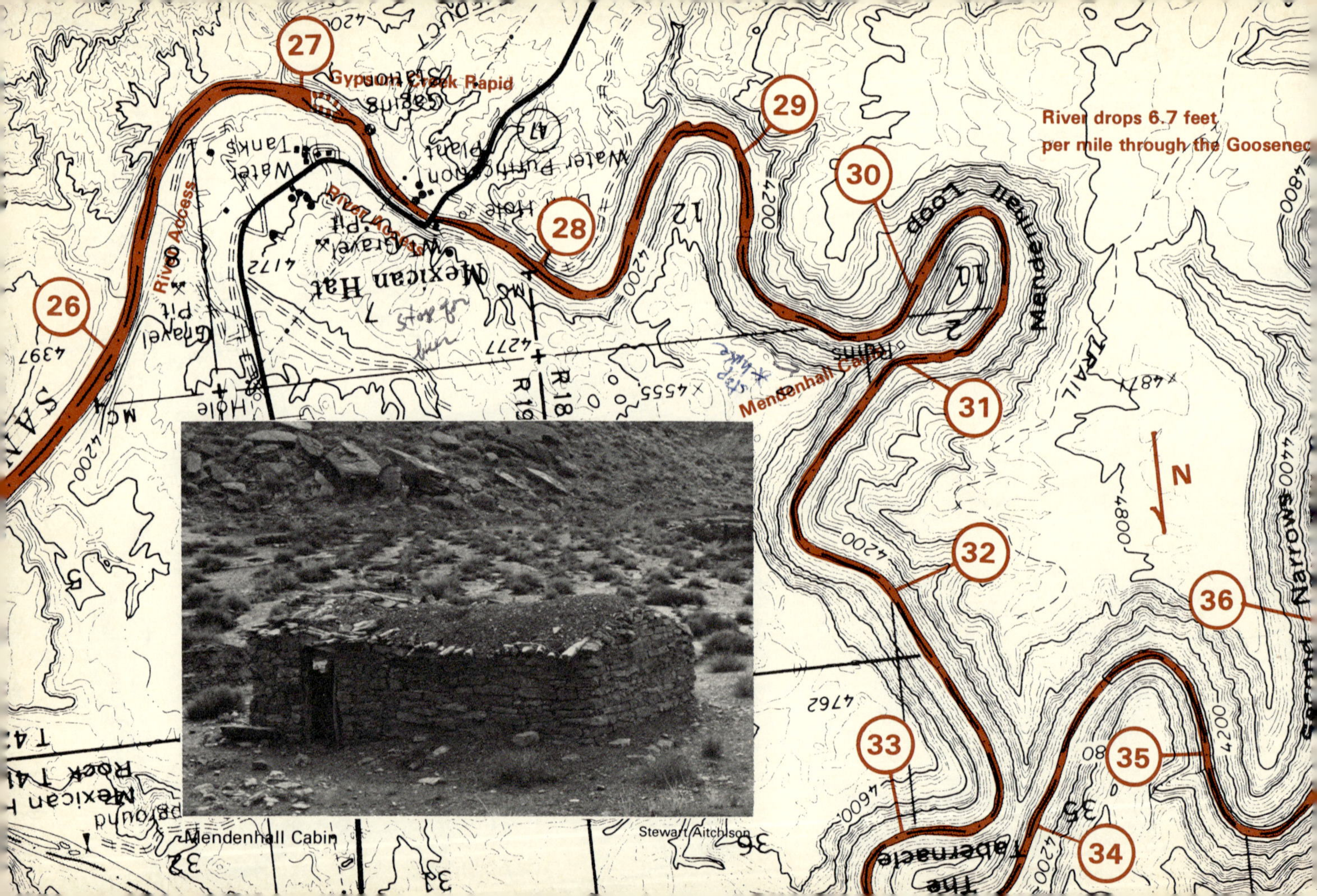

Mendenhall Cabin

Stewart Aitchison

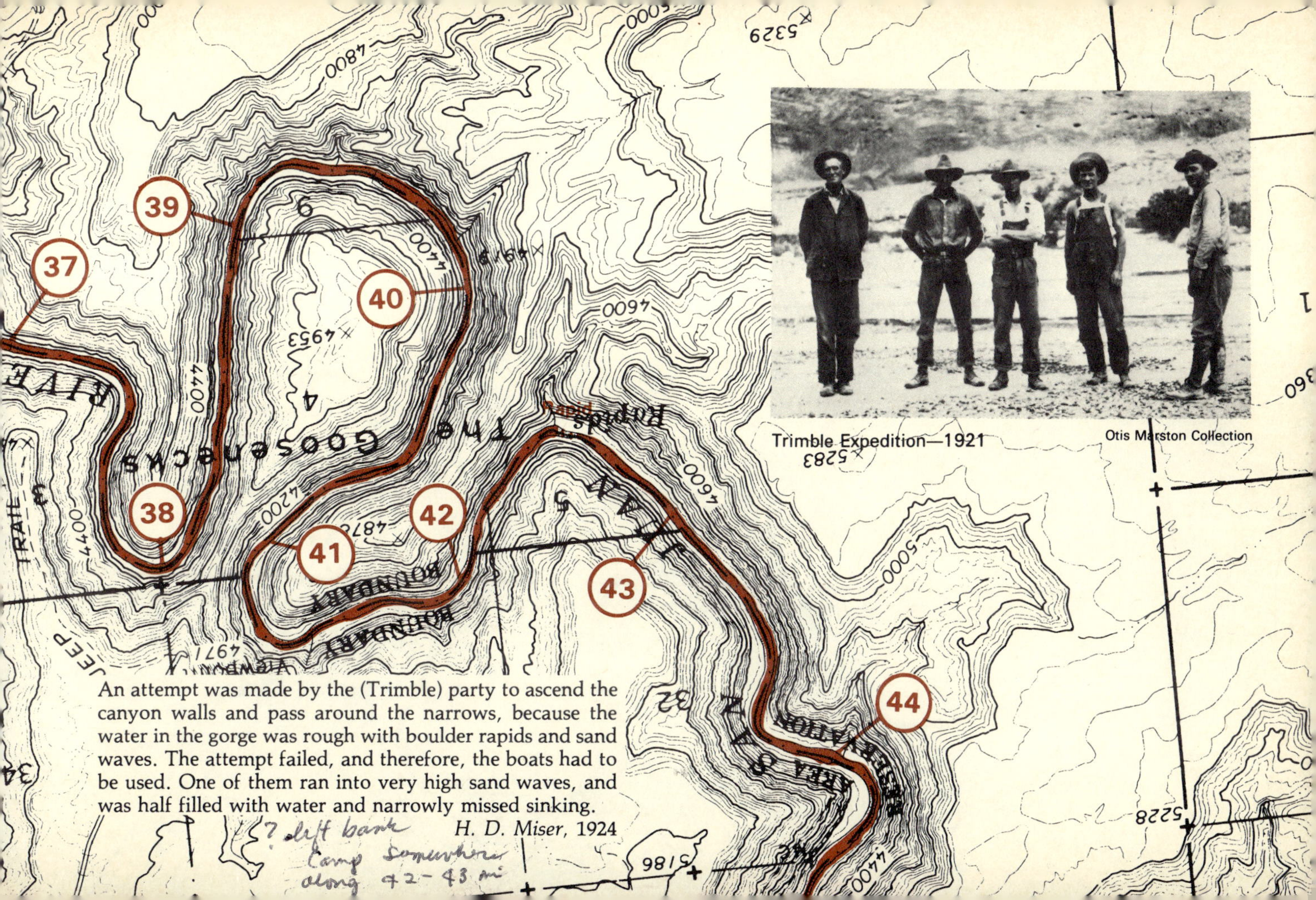

Trimble Expedition—1921

Otis Marston Collection

An attempt was made by the (Trimble) party to ascend the canyon walls and pass around the narrows, because the water in the gorge was rough with boulder rapids and sand waves. The attempt failed, and therefore, the boats had to be used. One of them ran into very high sand waves, and was half filled with water and narrowly missed sinking.

H. D. Miser, 1924

The (Honaker) trail was built (in 1904 by gold prospectors) with the intention of using pack animals on it, but the only horse ever to attempt the descent fell off from a particularly steep, narrow stretch known as "The Horn" and its bleached bones may still be seen lying at the base of the cliff. The canyon wall descended by the trail is only 1235 feet high, yet the trail is so crooked that it is about 2½ miles long.

H. D. Miser, 1924

H.D. Miser, U.S. Geological Survey

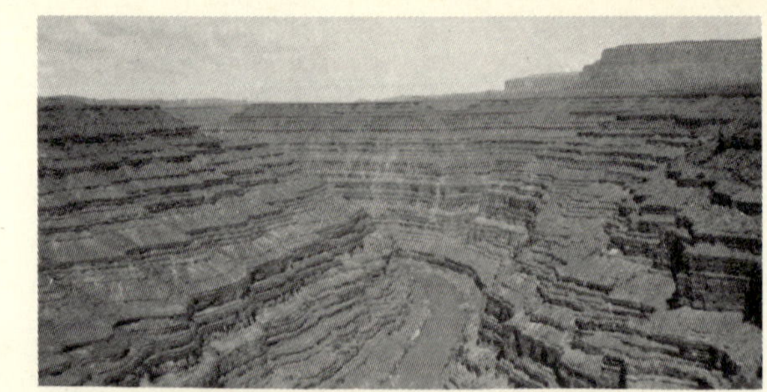

West of Honaker Trail

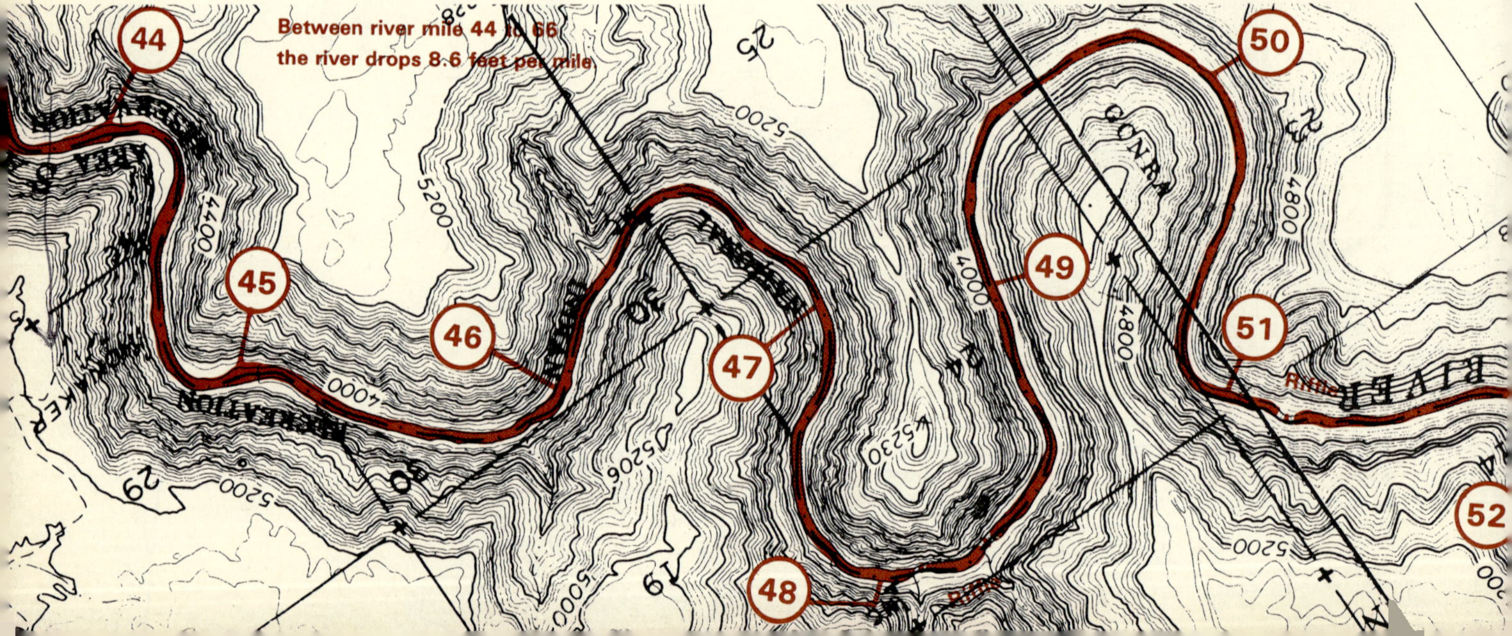
Between river mile 44 to 66
the river drops 8.6 feet per mile.
44
45
46
47
48
49
50
51
52
5200
4400
4000
4800
5230
Riffle
RIVER

Prospecting for gold in San Juan Canyon began in 1892, when tales of fabulously rich deposits caused 1,200 men to stampede to the canyon. After spending a few months there, they went away empty handed but prospecting continued at a few places as late as 1915. The gold is extremely fine in grain, being known as flour gold, and occurs mostly in placer deposits along the river channel and terrace gravels.

H. D. Miser, 1924

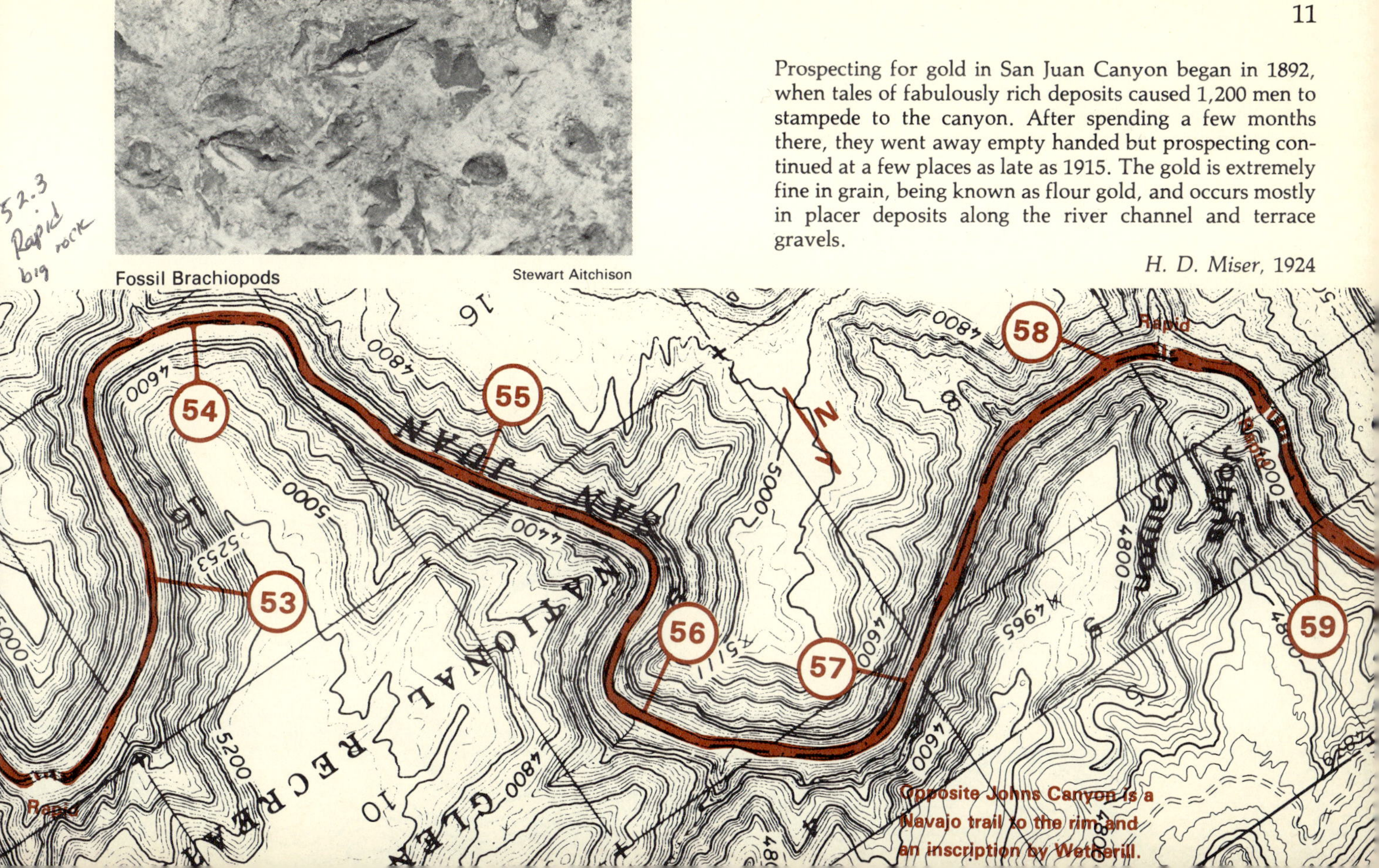

Fossil Brachiopods

Stewart Aitchison

camped above Govt Rapid

The loaded boats were run through a small rapid half a mile above the mouth of John's Canyon, but one of the boats containing two members of the party not only narrowly missed striking the canyon wall but struck a boulder and was burst on one side from bow to stern. The boat was nearly filled with water by the time a landing was reached. Then the wet equipment was unloaded and the boat was dragged ashore and repaired.

H. D. Miser, 1924

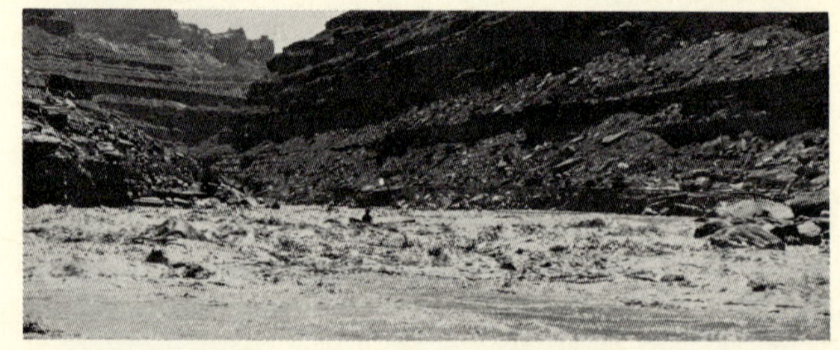

Government Rapid

H.D. Miser, U.S. Geological Survey

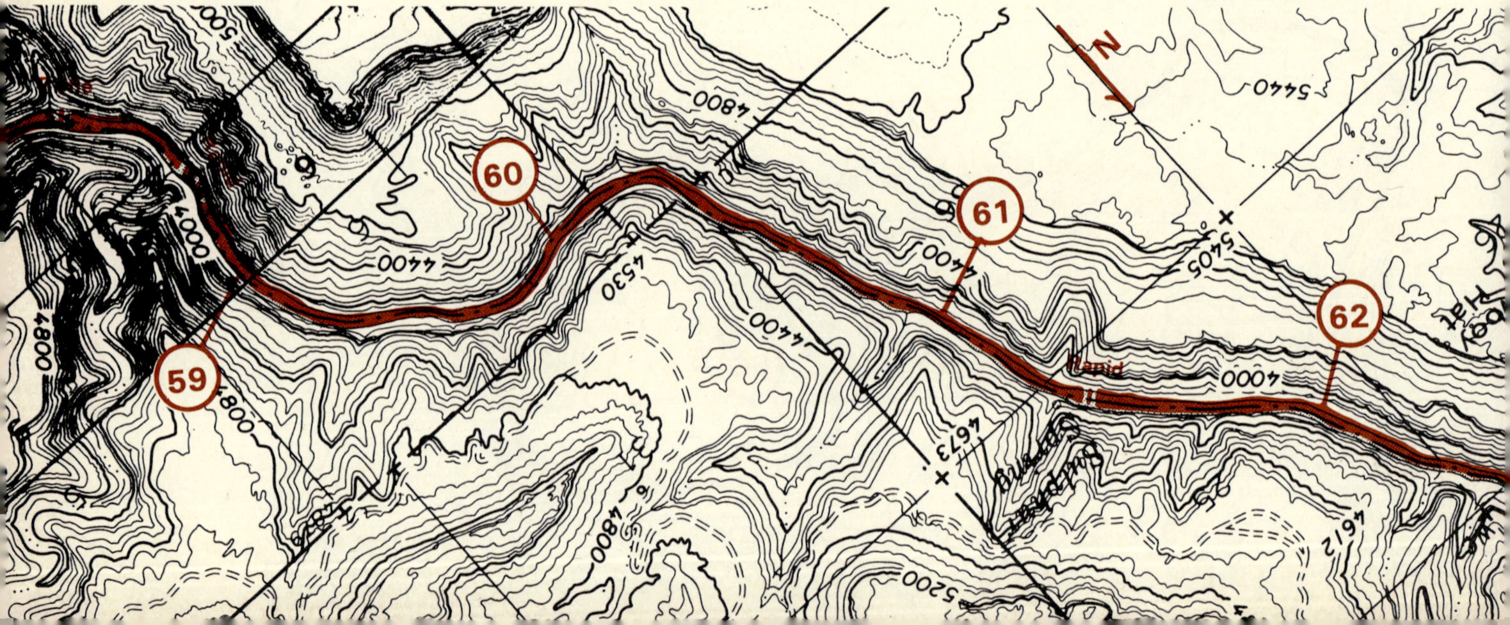

camped at Slickhorn Fri nite

Stewart Aitchison

Drilling Rig at Slickhorn

A round jar having the form of an olla was found beneath a narrow projecting ledge on the right bank of the river about one mile above the mouth of Slickhorn Gulch. In this vicinity oil seeps occur in the river and at the water's edge for a distance of a mile and a quarter. Perhaps the cliff dwellers who made pottery of this type came to the locality to obtain the oil for fuel or medicine.

H. D. Miser, 1924

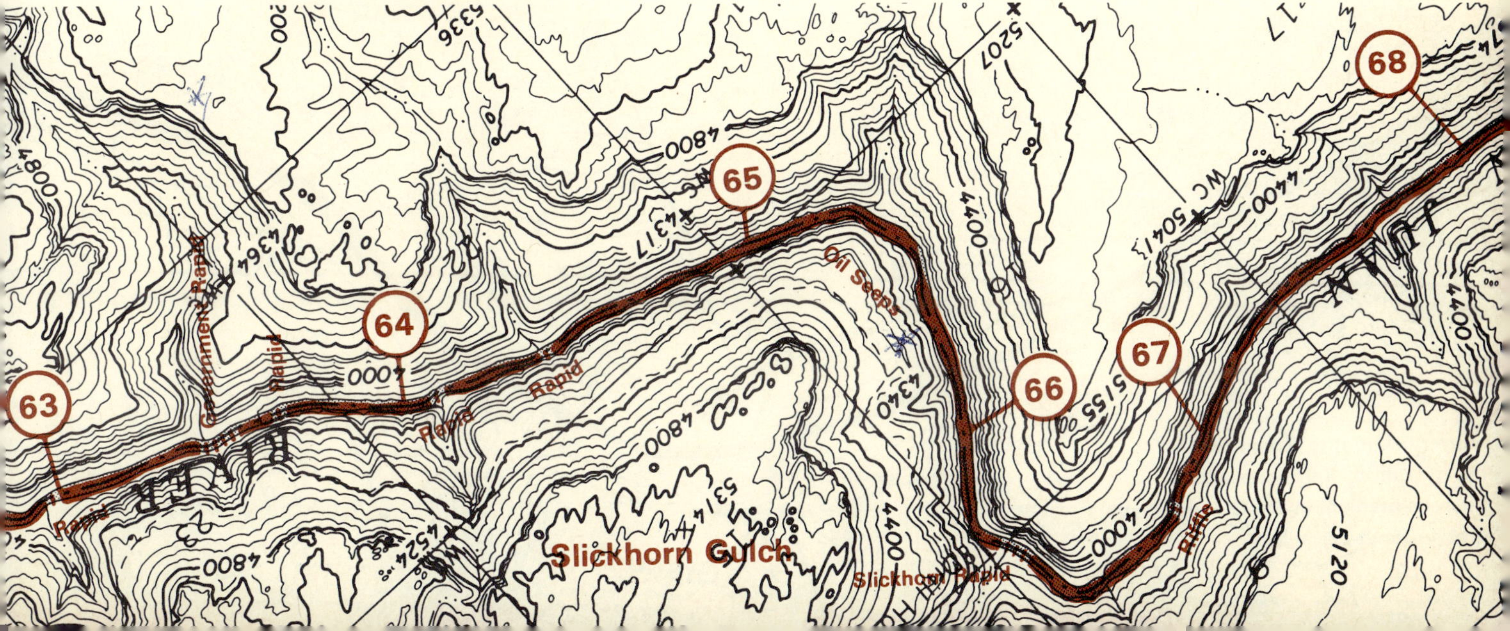

63 64 65 66 67 68

Government Rapid

Rapid

Oil Seeps

Slickhorn Gulch

Slickhorn Rapid

Riffle

RIVER

JUAN

Grand Gulch drains nearly all the territory southwest of the Elk Mountain from McComb Wash to the Clay Hills, about 1000 square miles of territory. It is the most tortuous cañon in the whole of the Southwest, making bends from 200 to 600 yards apart almost its entire length, or for 50 miles; and each bend means a cave or overhanging cliff. All of these with an exposure to the sun had been occupied by either cliff houses or as burial places.

Richard Wetherill, 1893

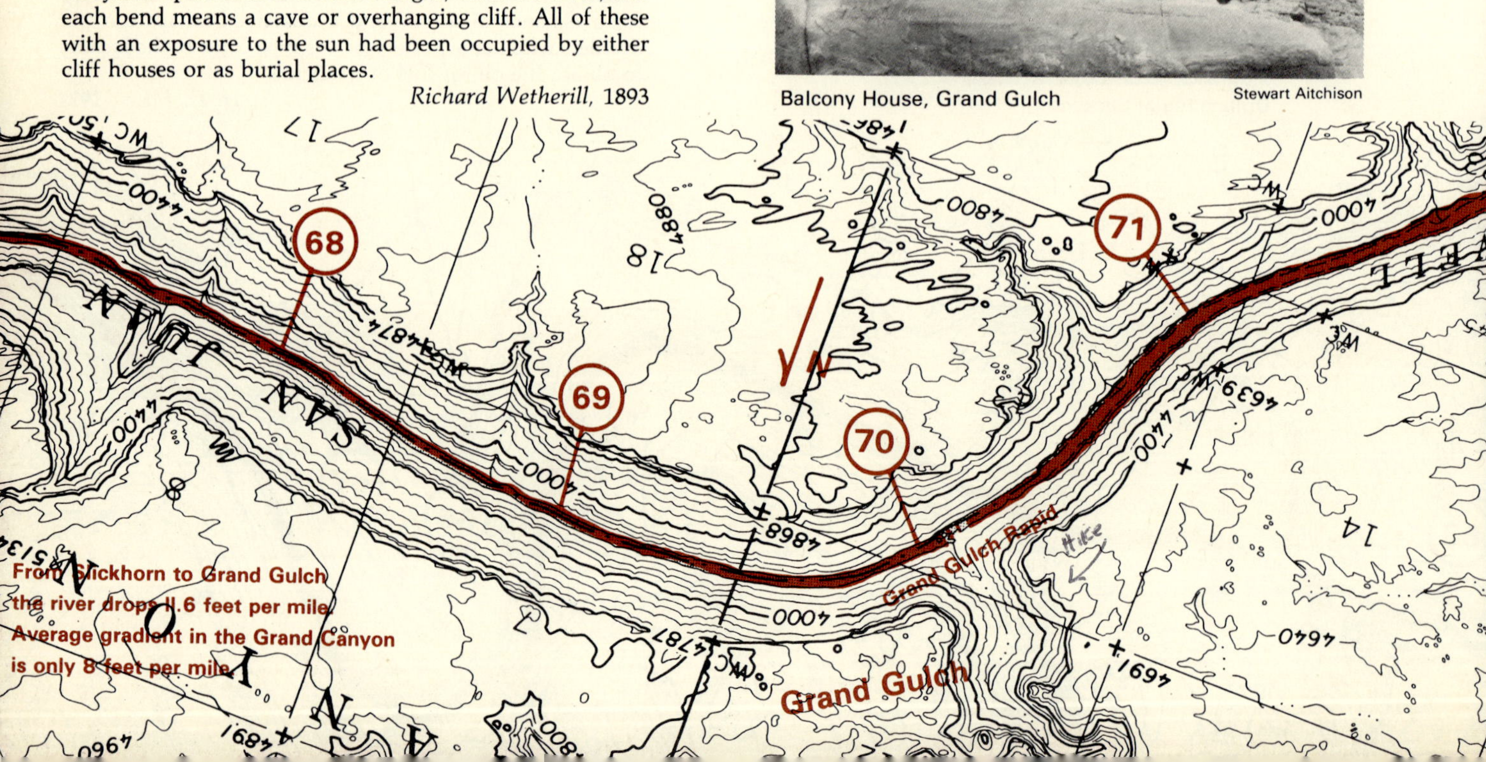

Balcony House, Grand Gulch

Stewart Aitchison

A.H. Jones

Oljeto Trading Post, 1909

Last campsite (L) bank

In 1906, after being granted permission from the Navajo chief, Hoskinini, to stay at Moonlight Water (Oljeto), John and Louisa Wetherill established one of the first trading posts in Navajo country. Like his brothers, John was fascinated by the Anasazi ruins and during his explorations near the San Juan discovered many archaeological sites as well as Rainbow Natural Bridge.

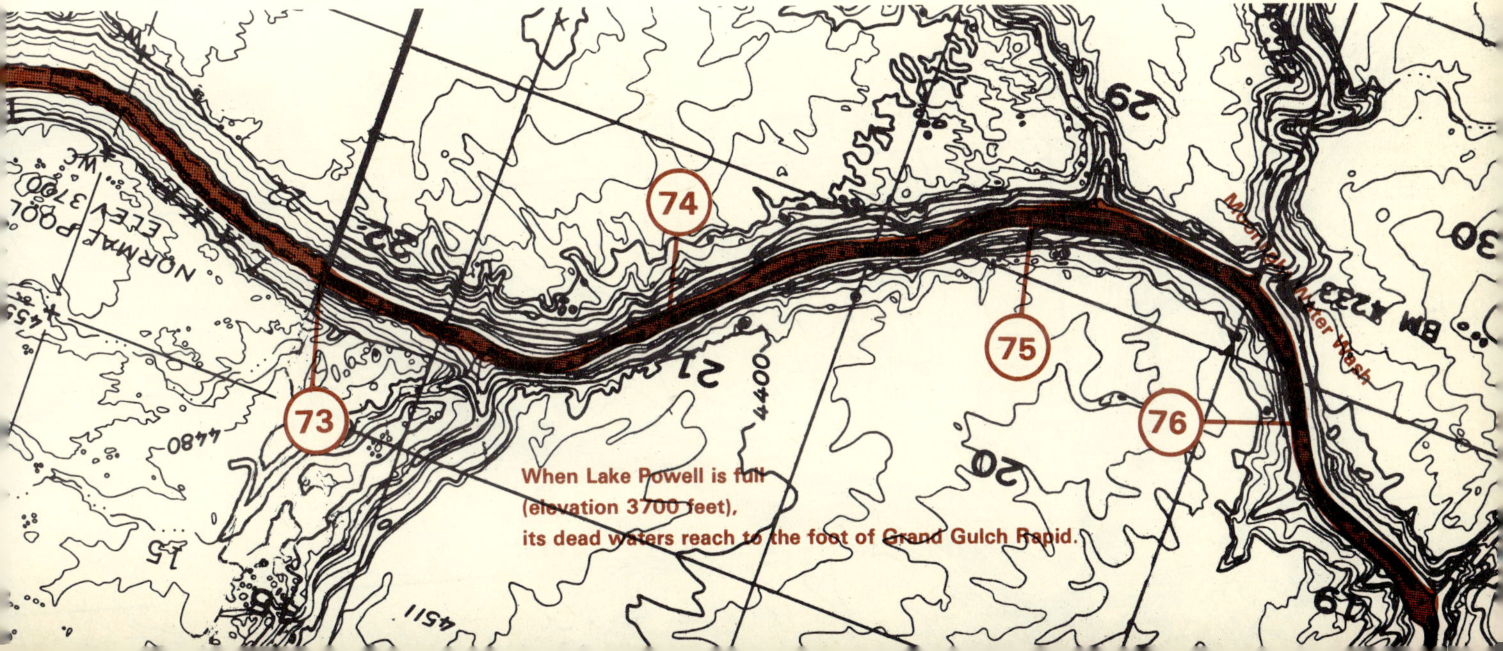

74

73

75

76

Moonlight Water Wash

NORMAL POOL ELEV 3700

BM 4233

4400

4480

When Lake Powell is full
(elevation 3700 feet),
its dead waters reach to the foot of Grand Gulch Rapid.

In Whirlwind Draw,

shallow depressions with pools contain dozens of tadpole shrimp that look for all the world like miniature horseshoe crabs, with a shield-shaped carapace that covers the head and thorax, and at the ends of each tail two thin spiny projections. They breathe through delicate paddlelike appendages. . . . The rhythmic fluttering of these legs is constant, sometimes stirring up . . . algae, bacteria, protozoa, whatever is small enough to ingest.

Ann Zwinger, 1978

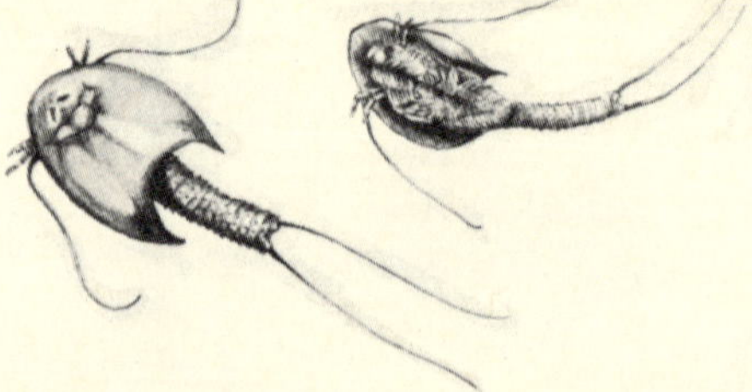

Tadpole Shrimp

Ann Zwinger

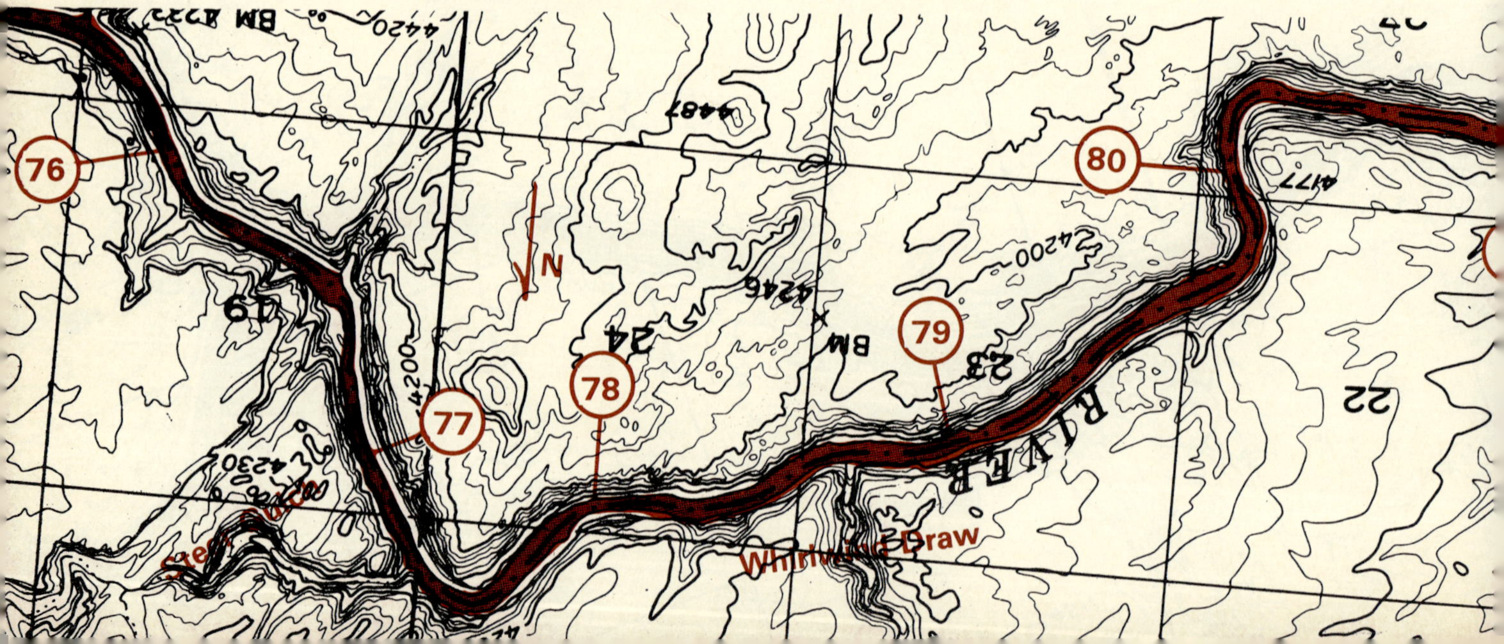

San Juan River Canyon — David Maren

And, as it rose in rock, so it ends in rock . . . sediments from more ancient mountains, layered, worked, reworked, laid in witness layers.

A cliff swallow makes a last sweep of the sky, then shafts like an arrow downstream. The wind stills. It is too quiet. I do not want to hear the river ending.

Ann Zwinger, 1975

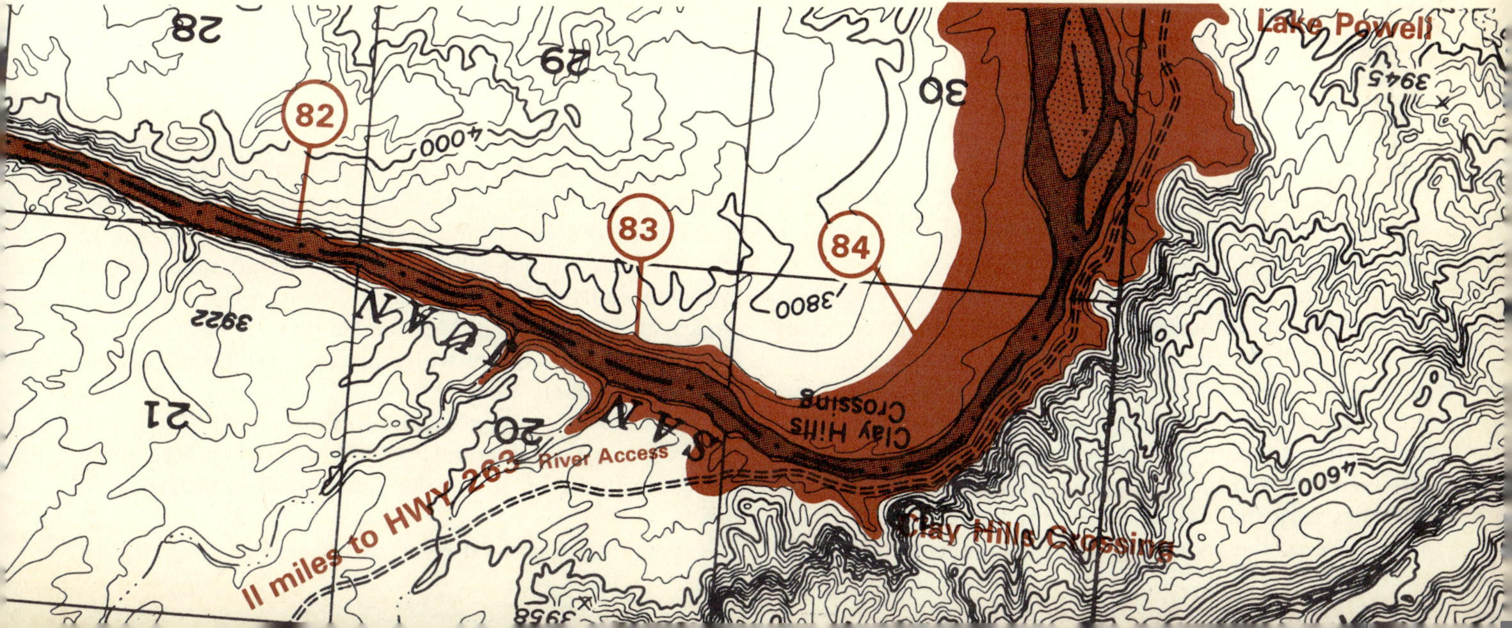

Lake Powell
3945
82
4000
83
84
3800
River Access
Il miles to HWY 263
4600

RIVER TRIPS

On your own. What the San Juan River lacks in huge whitewater (the largest rapid is a Class III), it easily compensates with spectacular scenery and fascinating prehistoric and historic sites.

The river is usually floated in small rafts, kayaks, sportyaks, dories, or occasionally in open canoes. The particular craft used is primarily a function of water flows. In the past the river's volume was controlled by the amount of snow melting in the San Juan Mountains and rain showers over the watershed. The river could fluctuate tremendously; for instance, in 1905 the flow varied from a low of 40 cubic feet/second to over 24,000 cubic feet/second. In 1911, the river rose nearly 50 feet above normal and washed away the bridge at Mexican Hat.

Today the San Juan's flow is regulated by Navajo Dam in New Mexico although thundershowers below the dam may still occasionally cause significant flooding. The dam supposedly never releases less than 500 cubic feet/second so in theory a small boat could always get down the river. Remember, though, that the river is braided into several channels between Montezuma Creek and Comb Wash and becomes very shallow, at times only inches deep.

Because of the river's steep gradient, the water moves surprisingly fast. There is always the danger of being swept into a snag, cliff, or rock. Stay alert!

On about 1,000 cubic feet/second, the trip from Montezuma Creek to Sand Island takes 5 to 10 hours, Sand Island to Mexican Hat takes one long or two short days, and Mexican Hat to Clay Hills Crossing is a three to four day trip. If you plan to explore the side canyons, allow yourself more time. Also strong upstream winds, common in spring and during unsettled weather, may significantly slow your downstream progress. Remember, too, that there is usually no river current from Grand Gulch to Clay Hills Crossing, some 13.5 miles, because of Lake Powell.

If Lake Powell has recently been drawn down, you may have current all the way to the Clay Hills take-out but there may also be a mud-flat to cross to reach your vehicle.

The car shuttle distances are (1) Montezuma Creek to Sand Island—about 50 paved miles *or* 20 miles, 15 of which are dirt; (2) Sand Island to Mexican Hat—about 24 miles, all paved; and (3) Mexican Hat to Clay Hills Crossing —about 74 miles, 11 of which are dirt. See Location Map for road numbers.

If you wish to hire someone to do the car shuttle, you may contact Valle Trading Post in Mexican Hat, Recapture Lodge in Bluff, or one of the commercial outfitters listed below.

Other than the usual river paraphernalia and required

equipment, which is outlined by the Bureau of Land Management (BLM) when you obtain a permit, it is a good idea to carry your own drinking water. The river water is quite silty and side streams and springs are unreliable. A PERMIT IS REQUIRED. These are available without cost from the Bureau of Land Management. This federal agency administers boating on the San Juan from Montezuma Creek to Clay Hills Crossing. Also if you plan to hike more than 4 miles from the river up either Slickhorn or Grand Gulch, a BLM hiking permit is required. They can supply you with current information concerning required gear and water flow data. Their address is: Bureau of Land Management P.O. Box 7, Monticello, UT 84535, (801) 587-2201.

Furthermore, the south side of the river is Navajo Indian Reservation and a permit is necessary for camping or hiking on their land between Montezuma Creek and Chinle Wash. Their address is: Navajo Tribal Council, Parks and Recreation Dept., Window Rock, AZ 86515, (602) 871-5561.

> The traveler should never leave camp without a supply of water and should keep in mind the deceptive character of mirages.
>
> *H. E. Gregory*, 1916

Commercial Outfitters

Canon Tours P.O. Box 1386 Durango, CO 81301	Offers raft trips.
Fastwater Expeditions P.O. Box 365 Boulder City, NV 89005	Offers sportyak trips.
Grand Canyon Youth Expeditions, Inc. Route 4, Box 755 Flagstaff, AZ 86001	Offers raft and kayak trips.
Holiday River Expeditions, Inc. 519 Malibu Avenue Salt Lake City, UT 84107	Offers raft, sportyak, and inflatable kayak trips.
Ken Sleight Expeditions P.O. Box 338 Green River, UT 84525	Offers raft trips.
O.A.R.S. P.O. Box 67 Angels Camp, CA 95222	Offers raft trips.
Outlaw Trails, Inc. P.O. Box 336 Green River, UT 84525	Offers raft and sportyak trips.
San Juan Expeditions P.O. Box 1206 Moab, UT 84532	Offers raft trips.

Tag-A-Long Tours P.O. Box 1206 Moab, UT 84532	Offers raft and sportyak trips.
Wild and Scenic, Inc. P.O. Box 460 Flagstaff, AZ 86002	Offers sportyak and raft trips.
Wild Rivers Expeditions P.O. Box 118 Bluff, UT 84512	Offers raft trips.

On the River — Stewart Aitchison

GEOLOGY

When you first come to the San Juan country, many questions come to mind. Why are the rocks red? Why are they layered? Why are some layers tilted? Why do some rocks form cliffs while others form slopes? Has the river always been where it is today? How did the entrenched Goosenecks form?

The rock layers that make up the cliffs of the San Juan canyons were originally horizontal. The rocks are a record of seas, rivers, and sand dunes that covered the San Juan country millions of years ago. To give some perspective, the earth is 4,600 million years old. The oldest hard-shelled fossils are about 600 million years old. The oldest rocks exposed in the San Juan canyons are about 320 million years old. The San Juan River is at most 25 million years old, and man, as a species, is less than 2 million years old.

What do rocks tell a geologist about past environments? Geologists look at rocks being formed today and assume that the same processes formed similar rocks millions of years ago.

For instance, the rocks that make up the canyon walls in the Goosenecks are made of calcium carbonate and contain fossil corals. Today very similar rocks are forming off the coast of Florida. The sea contains trillions of tiny plants and animals with calcium carbonate shells. Coral reefs grow in warm shallow seawater. When the coral dies, its calcium carbonate shell is buried by more shells eventually forming a fossil reef or biohermal bank. By analogy, the rocks of the Goosenecks must have formed in a similar fashion.

How do we know the age of the fossil bioherms? Geologists recognized that when a layer of limestone (or any sediment) forms, it is younger than the layer below it. Using this principle, they realized that certain fossils were younger than others. Between 1760 and 1891 a relative geologic time scale was established using fossils. Later the fossiliferous limestones of the Goosenecks were found to contain fossils similar to limestones studied in Pennsylvania and were thus called Pennsylvanian in age.

It was not until the early 1900s that radioactive decay rates were used to determine the actual age, in years, of certain minerals within rocks. By comparing radiometric dates with fossils, geologists have decided that the Gooseneck limestones are 295 to 320 million years old. These limestones are now over 4,000 feet above sea level and, as will be seen later, they were pushed up and tilted starting 60 million years ago (MYA) at the same time that the Rocky Mountains formed.

Not all of the rocks along the San Juan River were formed as marine deposits. Many of the slope-forming red rocks

AGE Millions of years ago	TIME PERIOD	
140 to 195	JURASSIC	Morrison Formation
		Bluff Sandstone
		Summerville Formation
		Entrada Sandstone
		Carmel Sandstone
		Navajo Sandstone
195 to 230	TRIASSIC	Kayenta Formation
		Wingate Sandstone
		Chinle Formation
		Moenkopi Formation
230 to 295	PERMIAN	De Chelly Sandstone
		Organ Rock Shale
		Cedar Mesa Sandstone
		Halgaito Shale
295 to 310	PENNSYLVANIAN	Honaker Trail Formation
		Paradox Formation

1000
FEET
500
250
0

Geologic Column

were deposited by rivers. By studying present day Colorado River sediment near the U.S.–Mexico border, geologists have found that the red color is due to "rusting" of iron-bearing minerals in the sand and silt. Fossils of land plants associated with floodplains also indicate a "fluvial" or river origin for these layers.

How could limestone deposited on the seafloor be buried by river sand? Either the sea level dropped or the seafloor rose. Geologists believe that the earth's crust is made of large plates that move slowly (inches per year) relative to each other. Where one plate is colliding with another, mountains form and sediments deposited below sea level are uplifted. For example, the Appalachian Mountains were formed when Africa and Europe "collided" with eastern North America about 300 MYA. When this occurred, limestones from the "proto-Atlantic" seafloor were elevated above sea level and buried by river sands.

In order for a layer of sediment to be preserved in the geologic record, it must be buried by more sediment. Downwarping of the sedimentary layer creates a basin in which more sediment can collect. Elevation of an adjacent area can provide sediment through erosion to bury the basin deposits. These processes have alternated throughout geologic time in the San Juan canyons area.

Geologists divide rocks into units called formations. Generally speaking, a formation represents a rock layer formed in an environment that is distinguishable from the environment of deposition above and below that layer. A formation may consist of more than one rock type (for instance, sandstone and shale) and may or may not contain fossils. Formations are often bounded by unconformities which represent periods of erosion or nondeposition. However, formation boundaries may be gradational. These are arbitrarily picked and difficult to locate. Formations also have lateral boundaries. This means a formation can "pinch out" or grade laterally into another rock type. Imagine a sand dune area adjacent to a river floodplain. The future result may be two formations, one of windblown sand, the other a fluvial sandstone, side by side and of the same age.

Each of the formations that appear along the San Juan River was deposited over thousands of square miles. Most are limited to the Colorado Plateau of eastern Utah, northeastern Arizona, northwestern New Mexico, and western

Monument Valley

Stewart Aitchison

Colorado. This unique region is characterized by multi-colored layered rocks forming high plateaus that have been dissected into canyons by the Colorado River and its tributaries. Thus, the events told by the San Juan canyon rocks are really part of a much larger regional story.

Early Pennsylvanian Time 310–320 MYA

The oldest rocks exposed in the San Juan River canyons are the fossiliferous limestones and siltstones that were deposited in a shallow marine environment. They are called the Paradox Formation.

Bioherms or fossil reefs in the Paradox Formation can be seen at river level from mile 15.5 to 17, where they appear as undulating layers of limestone. Bioherms are also exposed in the massive limestone cliffs in the Goosenecks and near the Honaker Trail, mile 44. Fossils found in the bioherms include coral, brachiopods, and bryozoa but most of the fossils are of green algae.

Silty gypsum beds can be seen from mile 18 to 19. Most of the gypsum has been leached out causing the overlying layers to slump into the canyon.

The yellow layer at mile 14.4 is near the top of the Paradox Formation. Oil located in fossil bioherms is pumped from Paradox rocks in the Aneth oil field east of Bluff. Paradox Formation salt is mined near Moab for potash.

Late Pennsylvanian Time 295–310 MYA

The Honaker Trail Formation overlies the Paradox. It consists of fossiliferous sandstone, siltstone, shale, and limestone that were deposited in a shallow sea. These rocks form the high canyon walls in the deeper canyons. Due to uplifting, tilting, and folding that began 60 MYA, the Honaker Trail Formation appears at river level at mile 9.3, disappears at mile 20.2, reappears at river level just above Mexican Hat Bridge and disappears again at mile 72.

Corals, clams, brachiopods, bryozoa, and crinoid stems are abundant. A good place to see these fossils is in the bed of the small canyon on the right at mile 12.9.

Oil in the Mexican Hat oil field is pumped from the Honaker Trail Formation. The oil was derived from marine plants and animals that were rapidly buried by sand and silt, trapping the organic material. Eventually thousands of feet of sediment covered the organic remains; the pressure of those sediments and the resultant heat converted the organic material to oil. The oil then seeped into pore spaces within the sandstone and limestone layers.

Early Permian Time 270–295 MYA

The ancestral Rocky Mountains (also called the Uncompahgre Uplift) had become a major range by Permian time. This uplift was probably due to the collision of the South

American and southern North American plates. Streams draining these mountains deposited gravel and coarse sand near the range (Cutler Formation) and red silt and mud on the floodplains (Halgaito Shale) in the San Juan canyons area.

The Halgaito Shale Formation forms the red hogbacks or ridges on the right side of the river from mile 7.5 to 9 and Mexican Hat Rock at mile 23.5. It is also exposed from mile 70 to 70.2. Plant and reptile fossils can occasionally be found in it.

White sand carried by ocean currents from the northwest was deposited along a fluctuating shoreline and covered the Halgaito. Sometimes the sea level dropped and the white sands were reworked by the wind. Thus the resulting Cedar Mesa Formation is partly marine sandstone and partly eolian. This is similar to events now occurring along the Gulf Coast near Galveston, Texas.

In the valley below Comb Ridge, the Cedar Mesa Sandstone grades eastward into a red silt deposited in a lagoonal environment; this can be seen at mile 9.2. At this location, the Cedar Mesa is lighter red than the underlying Halgaito and the overlying Organ Rock Shale and contains gray-green gypsum beds.

Good exposures of crossbedded marine Cedar Mesa Sandstone can be seen along the river from mile 70.2 to Clay Hills Crossing.

Lime Ridge Anticline — H.D. Miser, U.S. Geological Survey

Mid Permian Time 230–270 MYA

The Organ Rock Shale overlies the Cedar Mesa Sandstone. Like the Halgaito Shale, the Organ Rock represents a river floodplain, changing to tidal flat deposits farther west. There are thin exposures of Organ Rock Shale on the trail to the Mule Ear Diatreme at river mile 9.1, but the best exposures are in the red slopes forming the base of the monuments in Monument Valley.

Steeply dipping "cross beds" in the orange-brown De Chelly Sandstone indicate that strong northeast winds picked up red sand from the Cutler Formation and deposited it on the Organ Rock Shale. The De Chelly Sandstone is thickest in Canyon De Chelly National Monument and pinches out in the San Juan canyons area. It is present as a low ridge along the river from mile 7.3 to its intersection with the Mule Ear Diatreme at mile 9.1 and is not present at Clay Hills Crossing. The sheer cliffs of the monuments in Monument Valley consist of De Chelly Sandstone.

Late Triassic Time 220–215 MYA

In Early Triassic time, subsidence of the land surface to sea level allowed deposition of extensive tidal and river floodplain sediments. This is the Moenkopi Formation. The red silts were derived from the eroding ancestral Rocky Mountains and Cutler Formation and carried westward by sluggish streams. The Moenkopi is exposed in the long valley north of the Mule Ear Diatreme.

In Late Triassic time, renewed uplift of the ancestral Rockies and extensive volcanic activity in southern Arizona provided large amounts of sediment to the rivers that deposited the Chinle Formation. The upper layers of Chinle represent wide shallow lakes that formed as the mountains were eroded to lower elevations. The green-gray color of some Chinle layers below the Mule Ear is due to clay derived from volcanic ash carried from the volcanic highlands in southern Arizona. The Chinle Formation forms the variegated landscapes of the Painted Desert and Petrified National Park in northern Arizona.

The massive orange cliff that crosses the river at mile 7.0 and forms the prominent Mule Ear to the south is the Wingate Sandstone. Winds from the north carried large volumes of sand into southeastern Utah burying the Chinle shales. The Wingate is not exposed at river level again but can be seen capping the high cliffs to the west of Clay Hills Crossing.

The Kayenta Formation is primarily a fluvial sandstone. It was deposited by rivers flowing southwest over the Wingate sands. The small cabin on the right at mile 6.4 is on a ledge in the Kayenta Formation. The earliest mammals are thought to have evolved in Late Triassic time.

Late Triassic to Early Jurassic Time 190–200 MYA

Starting in Late Triassic time and continuing into the Jurassic period, strong winds from the north created a sand dune field that stretched from southern Wyoming to northern Arizona and western Colorado to eastern Nevada. The resulting formation is the Navajo Sandstone. Westernmost portions may have been deposited by marine currents but the massive, cross-bedded exposures forming cliffs from Sand Island to mile 5.6 indicate an eolian origin. Scattered fossil playas (intermittent desert lake deposits) suggest that the Navajo was deposited near groundwater level or possibly near sea level.

The dark stains streaking the Navajo Sandstone cliffs are called desert varnish; this is an oxidized mineral patina. Dinosaur tracks have been found in the Navajo Sandstone and Kayenta Formation. The Navajo, Kayenta, and Wingate form Comb Ridge, mile 7–8. These same formations are exposed in Zion National Park and the cliffs surrounding Lake Powell.

Mexican Hat Rock H.D. Miser, U.S. Geological Survey

Mid Jurassic Time 160–175 MYA

Red-brown mud, silt, and sand were deposited on a slowly sinking seafloor throughout Utah during Mid Jurassic time. The resulting rocks, called the Carmel Formation, erode to form the red ledgy slope above the Navajo Sandstone in the Sand Island area.

The orange-brown cliff overlying the Carmel is the Entrada Sandstone. In the San Juan canyons area, it represents a marine sand deposit but farther east and north, it is eolian. The arches in Arches National Park are in the Entrada.

The Summerville Formation forms the low red ledgy cliff along the north side of Bluff. It consists of calcareous sandstone and siltstone deposited in tidal flats. A local beach sand later spread out over the Summerville in the Bluff area leaving a thick layer of light brown sand known as the Bluff Sandstone. This sandstone forms the light colored cliff on the skyline north and south of the town of Bluff.

During the latest Jurassic times, a volcanic highland in the California-Nevada area becomes the continental divide. Streams draining this mountain range (the Sierra Nevada–Sevier Highlands) carried volcanic ash as far east as Nebraska, burying Utah and Colorado with the Morrison Formation. It is in these rocks that dinosaur fossils are preserved in Dinosaur National Monument. The Morrison is exposed north and east of Bluff, above the Bluff Sandstone cliffs on the eastern skyline viewed from Sand Island.

Late Cretaceous Time 65–95 MYA

During Early Cretaceous time, the San Juan canyons area was above sea level. It was either undergoing erosion or fluvial and coastal plain sediments were being deposited. By Late Cretaceous time, a major seaway covered all of the central United States, at times extending into eastern Utah. East-flowing rivers draining the Sierra Nevada–Sevier Highlands deposited organic debris in coastal marshes from northern Arizona and New Mexico to central Wyoming. These deposits later became coal beds, many of which are being mined today.

Coal mining is a controversial issue in southern Utah because many national parks are located near large Cretaceous coal deposits. To reduce energy costs, power companies want to build coal-fired power plants near the mines. Many people are concerned that the mining activities and power plant emissions will mar the panoramic vistas that this region is famous for and contribute to the acid rain pollution problem.

Cretaceous rocks make up Black Mesa near Kayenta, the Book Cliffs in east-central Utah, and Mesa Verde near Cortez, Colorado.

Early Tertiary Time 40–65 MYA

The same forces that produced the Sierra Nevada– Sevier Highlands in Late Jurassic to Cretaceous time caused uplift of the Uinta and Rocky Mountains in Late Cretaceous to Early Tertiary time. West to east horizontal compression of the Colorado Plateau, caught between the Sevier Highlands and the Rocky Mountains, caused the earth layers to buckle into downwarps and upwarps. The Colorado Plateau surrounding the San Juan River area was uplifted and tilted toward the northeast.

The large downwarps became lakes that eventually filled with sediment brought in by rivers draining the surrounding mountains. The rocks that form the Pink Cliffs at Bryce Canyon National Park and the Roan Cliffs in Desolation Canyon of the Green River represent these Early Tertiary lake deposits.

Upwarps, such as the Kaibab, Circle Cliffs, and Monument, became highlands subject to erosion. When the Monument Upwarp formed, the originally flat lying strata of the San Juan canyons area were uplifted and gently folded.

Alhambra Rock Volcanic Neck H.D. Miser, U.S. Geological Survey

Early-Mid Tertiary Time 30–40 MYA

The eruption of a major volcanic system in the San Juan Mountains area of southwestern Colorado dominated the landscape 30 to 40 MYA. So much volcanic ash was produced that all but the higher peaks in the Rocky Mountains were buried. This resulted in a radial drainage system centered around the San Juan Mountains.

The ancestral San Juan River filled in the Blanding Basin east of the Monument Upwarp and eventually buried the south-central part of the upwarp. This enabled the ancestral San Juan River to meander over the Monument Upwarp and into the western Henry Basin, where the ancestral Colorado River was being impounded.

Uplift and northeast tilting of the Colorado Plateau continued. By mid Tertiary time the San Juan–Rocky Mountains had become the continental divide.

Late-Mid Tertiary Time 20–30 MYA

Several isolated laccolithic mountain ranges were formed on the eastern Colorado Plateau following the eruption of the San Juan Mountain volcanoes. Molten rock from deep in the earth was injected between sedimentary rock layers causing the overlying beds to dome up into mountains. Among these laccolithic ranges are the La Sals, the Abajos, the Henrys, and Navajo Mountain.

Numerous isolated volcanoes erupted on the Colorado Plateau during this time. Many of these consisted mostly of gas (steam) and pulverized rock—no lava or ash. The Mule Ear Diatreme at mile 9.1 is one of these. This "maar" type volcano erupted 28 MYA on a land surface a mile above the present day top of the diatremè. Large blocks of Jurassic and Cretaceous rocks (presently eroded from the immediate vicinity) that fell into the volcano's neck are found around the outer edges of the diatreme. Smaller blocks of rock from as much as 50 miles deep (the bottom of the earth's crust) were brought to the surface. Seventeen-hundred-million-year-old granite and rocks similar to those exposed in the bottom of the Grand Canyon can also be found in the diatreme. Look in ant hills on the diatreme for small red garnets.

By 20 MYA, the Colorado River had found its way to the Gulf of California. About 6 MYA, it began to carve the Grand Canyon, largely due to the northeast tilting and uplift of the Colorado Plateau. Most of the volcanic ash produced by the San Juan volcanoes had been eroded from the Rocky Mountains, and the rivers had begun to entrench themselves into the underlying rocks.

As the San Juan River eroded into the buried Monument Upwarp, its meanders became entrenched, forming the Goosenecks and other tight bends for which the river is famous. Some of these bends or loops were eventually cut off, leaving abandoned meanders. The river continued to cut deeper leaving the meander perched above the present river level. There is a perched meander at river mile 12.4.

Today—A Cross Section

In the last 20 MYA the Colorado Plateau has continued to rise and tilt northeast. Consequently, the rivers crossing it have carved deep canyons, the deepest being the Grand Canyon in the southwestern part of the Plateau. Similarly, the San Juan River is continuing to carve deeper into its canyons.

Not all of the geologic questions of the San Juan canyons have been answered. For instance, gravels similar to those in the present day river channel can be found in the terraces 100–200 feet above the river from Sand Island to the Mule Ear Diatreme, at the foot of Black Mesa (60 miles south in northern Arizona), and at isolated locations on White Mesa (south of Lake Powell).

The San Juan probably deposited the terrace gravels during the Ice Ages of ten thousand to a few hundred thousand years ago. However, if the San Juan deposited the gravels at Black and White mesas, it must have flowed south around the Monument Upwarp within the last few million years. Perhaps the neck of the Mule Ear Diatreme, with its greater resistance to erosion, forced the river to turn west across the Monument Upwarp. If this is true then the San Juan canyons across the Upwarp are only a few million years old instead of the 25 million generally accepted.

Sand terraces 30 feet above Chinle Wash, river mile 8.5, represent the creek bed a thousand years ago. Archaeologists believe the Anasazi Indians farmed the creek bottom after spring floods. A change in climate caused the creek to erode its own sediments leaving behind isolated terraces unsuitable for cultivation. This happened throughout the Southwest and may in part be related to the continued uplift of the Colorado Plateau which at present is an inch or two per year.

The San Juan River is well known for its sand waves. During flooding, these waves may have amplitudes of 10 feet. They only occur where the river has a sandy bottom and appear to be related to heavy silt loads. Careful observation indicates that sand waves are periodic, that is, they build up and die out on a regular cycle, usually no more than a minute at a time. The waves move slowly upstream, but this is not always obvious. Some of the largest sand waves occur just below Chinle Creek.

Most rapids in the canyons are formed by rocks brought in by side streams during flash floods. In a couple of cases, large boulders have fallen from the canyon wall forming a rapid.

As you float through the canyons of the San Juan River, try to imagine the ancient and remarkable environments that gave birth to the rocks. Feel the grit of the sandstone, the sharp edges on a fluted chunk of limestone; watch the power of the river pounding the rocks in a rapid; listen to the wind endlessly sculpturing the cliffs. The canyons seen today are but a short petrified moment in the history of the earth.

Natural Bridges National Monument W.T. Lee, U.S. Geological Survey

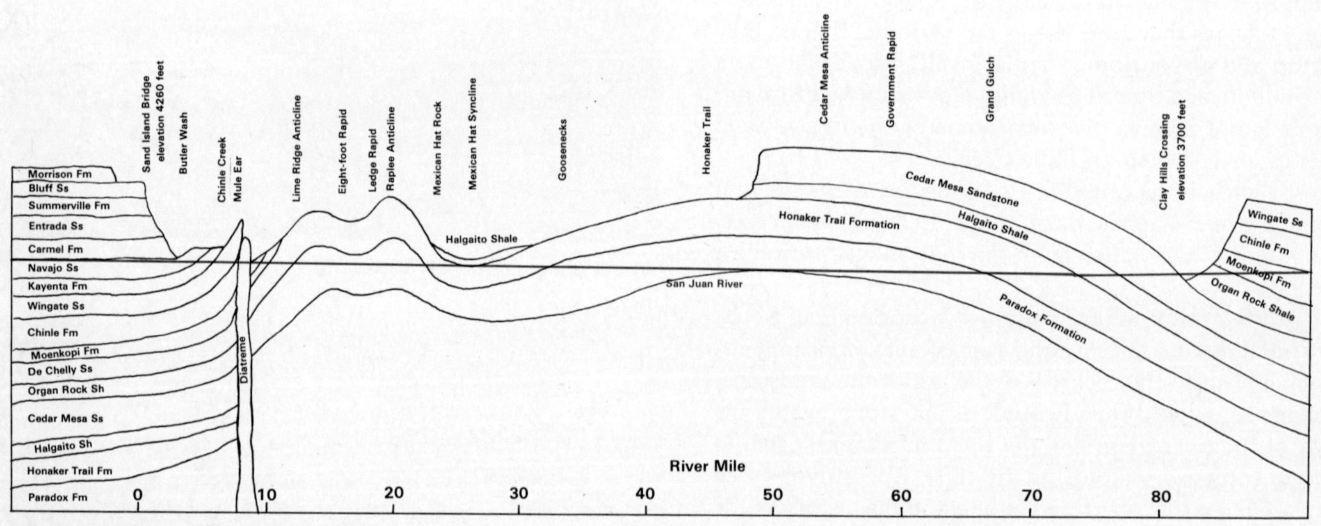

Geologic Cross Section

THE ANCIENT ONES

When the Navajo first ventured into the San Juan region some four hundred years ago, they found the remains of an old and extensive occupation. They called the former inhabitants "Anasazi," the "enemy ancestors."

The way of life evidenced by the Anasazi ruins represented a well developed stage in a long process of cultural adaptation to the lands of the Colorado Plateau. For many thousands of years the occupants of the Southwest had followed the seasonal availability of wild plants and animals, living in small seminomadic bands. Groups of these so-called Archaic hunters and gatherers were present throughout much of the San Juan drainage, although their numbers in southeastern Utah does not appear to have ever been large or their occupation continuous over long periods of prehistoric time.

The Archaic peoples built no permanent shelters and made no pottery. The few remnants they left behind include well made stone spear points and knives, other tools fashioned of animal bone, and the fascinating split twig figurines. These figurines were willow stems bent into small animal effigies and were probably used in magical ceremonies to ensure successful hunts.

Some three thousand years ago the spreading knowledge and practice of agriculture, based on developments which had occurred earlier in central Mexico and South America, began to bring slow but significant changes to the Southwest. The adoption of an agricultural lifeway allowed for the growth of more sedentary and larger populations, often in areas not as well suited for a foraging adaptation. The San Juan country proved itself a fertile land during much of the Anasazi period, which began nearly two thousand years ago.

For several hundred years the early Anasazi continued many Archaic ways, such as living in unmodified rockshelters or brush and pole pithouses. They grew some corn and squash but still relied heavily on hunting with nets, the rabbit stick, and spear. The gathering of wild plant foods was important as well. The Anasazi at this time, as had their Archaic predecessors, made fine baskets but no real pottery.

Around A.D. 500, the bean was incorporated into the Anasazi agriculture and diet. This event probably had a profound effect on the Anasazi life-style. Until this time, the agricultural part of their diet was poor in several essential nutrients, making hunting and gathering still necessary for survival. Most importantly, beans eaten along with corn provided a complete protein. Squash, in addition to providing its own nutrients, aided in the digestion of the bean. Freed from this reliance on the seasonal and scattered availability of wild plant and animal foods, the Anasazi could and did deepen their commitment to agricultural activities and rewards.

New knowledge and new resources undoubtedly brought many changes in the Anasazi way of life, although the changes were slow and perhaps imperceptible to the people themselves. Among these changes may have been a marked change in the ratios of youths, adults, and the aged in the population due to a more uniform diet, consistency of food supplies, and better developed food distribution systems, an old and time-honored form of social security. And with a less nomadic way of life, homes and storage structures were made more substantial and permanent with stone masonry construction. The harvests of corn, beans, squash, and several other crops were stored for the long winter months. Other changes included the introduction of cotton and the domestication of the turkey. Pottery making was introduced and refined and the spear and thrower was replaced by the more effective bow and arrow.

Increasing attention was given to a complex and formalized ceremonial expression of spiritual symbolism. This was, in part, in answer to the need to insure consistent rainfall and soil fertility, a need not strongly felt by the more broadly based nonagricultural Archaic peoples. A culmination of this development lives on in the modern

Anasazi Ruin — Stewart Aitchison

Pueblo Indian Kachina ceremonies.

After A.D. 1000, apartment-like complexes were being constructed and almost every settlement had at least one special ceremonial room called a kiva. Kivas were commonly circular and partially underground although at Turkey Pen Site in upper Grand Gulch there is an unusual square one perched on a ledge above the cave floor. Kivas served as social as well as ceremonial theaters and were an important focus of bonding through clan ties between dispersed Anasazi settlements.

The river terraces and rockshelters along the San Juan between Montezuma Creek and Chinle Wash contain numerous Anasazi sites in varying states of preservation. These sites date largely from the period A.D. 500 to A.D. 1300, with peak occupation apparently having occurred around A.D. 700–900. The groups that lived here were most closely related to those that occupied the Mesa Verde area at the same time. The sites include open terrace habitation sites, terrace gravel stone workshops, ledge granaries, rock art panels, sets of carved stone steps, and several well preserved rockshelter sites with living, storage, and ceremonial areas.

Further west on Cedar Mesa and in upper Grand Gulch and Slickhorn are a great many more remains of the Anasazi. These give evidence of three separate occupational peaks, A.D. 200–400, A.D. 650–700 and A.D. 1050–1275. These groups were also most closely related to the Mesa Verde Anasazi, with some evidence of ties with the Kayenta Anasazi who lived southward of Monument Valley in the broad region of north-central Arizona. The several granaries which can be seen along the San Juan below Grand Gulch were probably made by the later Anasazi occupants of the Cedar Mesa area.

Anasazi Flute Player David Maren

Anasazi pottery making, weaving and other arts as well as agricultural skills reached a point of considerable refinement by A.D. 1200. Climatic changes, arroyo cutting, resource depletion, and population overextension during the "good years" of A.D. 1000–1150, however, had already begun to threaten the local Anasazi settlements by the late 1100s. By A.D. 1300 the entire San Juan drainage had been abandoned. The San Juan Anasazi apparently moved out to join settlements that prospered in more favorable locations to become the modern Pueblo Indians of the Rio Grande Valley in New Mexico and the Hopi of the Little Colorado Valley in Arizona. The details of this exodus are lost in time, but evidence of relationships exists in artifact styles, cultural forms, and Pueblo myths.

After the withdrawal of the Anasazi, bands of Southern Paiute entered the San Juan country from the high plateaus of south-central Utah and the Great Basin farther west. These hunter-gatherers left little evidence of their occupation which was apparently rather sparse and which must have resembled in many ways that of the Archaic predecessors of the Anasazi. The Southern Paiute eventually established some permanent settlements and adopted limited agriculture at locations such as Paiute Farms on the lower San Juan and Moenave in the Little Colorado drainage. These settlements were absorbed or displaced by the Navajo who gained dominion over the region in the 1800s. Some of the descendants of these Southern Paiute live today at White Mesa Village north of Bluff. Broadly related groups in southwestern Colorado became known as the Ute and adopted a life-style enhanced by the mobility provided by the horse, introduced by European explorers.

The Navajo occupy much of the San Juan country today. Recently cultivated Navajo fields can be seen on the reservation land at the mouth of Chinle Wash. These and the many Navajo settlements in the wide country east and south of the river attest to a close and strong bond between these people and this land. The most striking cultural features along the San Juan itself, however, remain those of the ancient Anasazi. Although they are protected under state, federal, and tribal antiquities laws, time and increasing visitation have and are taking their toll on the condition and information potential of these sites. Please remember to treat these prehistoric places with respect by not climbing on walls or walking across easily eroded slopes. Take nothing from them.

CANYONEERS

The profusion of Spanish place names—Abajos, La Sals, Dolores, San Juan, and Mesa Verde—attests discovery and exploration of this region by Spain. One of the earliest expeditions to encounter the San Juan was led by two Spanish priests, Fray Francisco Atanasio Dominguez and Fray Silvestre Velez de Escalante.

In July 1776, they left Santa Fe, New Mexico, hoping to find a northern trail to Monterey in California. On August 5, they crossed the San Juan not far below its confluence with the Navajo River. They never made it to Monterey; early snowstorms eventually forced them to return to Santa Fe.

By the 1840s, U.S. trappers had possibly entered the San Juan canyons, but it was not until 1859 that the first official U.S. exploring party investigated the San Juan drainage. Captain J. N. Macomb was looking for a supply route between the Rio Grande Valley and the southern settlements of Utah. Macomb and his men were the first to record the location of the confluence of the Colorado (then called the Grand) and Green rivers deep in the canyonlands. The rugged slickrock wilderness inspired Macomb to write: "I cannot conceive of a more worthless and impracticable region." After exploring that area, the party "traveled south for about seventy miles, passing by the eastern base of the Sierra Abajo, until we struck the San Juan. . . on the 2d September 1859."

Bert Loper

Otis Marston Collection

Accompanying the expedition was the geologist J. S. Newberry. Newberry's report gave the world its first scientific, albeit cursory, glimpse at the canyonlands country of of southeastern Utah. From a campsite near the present townsite of Bluff, Newberry offered this description: "Nearly west from us a great gap opened in the high tablelands which limit the view in that direction; that through which the San Juan flows to its junction with the Colorado. The features presented by this remarkable gateway are among the most striking and impressive of any included in the scenery of the Colorado country." From this camp the Macomb party turned eastward traveling upstream and thereby avoided entering the great canyon of the San Juan River.

Although the canyons remained unknown for some years, several open areas along the river attracted the attention of settlers. Concern that this land might be invaded by others, the Church of Jesus Christ of Latter-day Saints (the Mormons) organized a mission to occupy this distant corner of Utah.

This mission is one of the most incredible sagas of personal faith and determination. A caravan of 83 wagons and 250 people—men, women, and children—rendezvoused near the town of Escalante in November 1879.

From there the wagon train headed eastward across the rolling sandy Escalante Desert to the rim of the nearly 2,000-foot deep-Glen Canyon. The terrain spread out before them was appalling. Stretching to the horizon was a land of bare sandstone thrown into a chaotic maze of cliffs and canyons.

With explosives, back-breaking labor, and remarkable ingenuity, a route was built down the canyon wall to the Colorado River. The wagons were driven down the steep trail and then floated across the river. More roadway was constructed up the other side. The Hole-in-the-Rock Expedition, as it became known, pushed onward.

A journey expected to take six weeks dragged into six months. Three babies were born en route. On April 6, 1880, with only 18 miles left to go to the original selected settlement site on Montezuma Creek, a decision was made in sheer exhaustion: "We'll stay here." And Bluff, Utah, was born.

During the summer of 1872, reports began to circulate that diamonds, along with rubies and sapphires, had been found in the West. One rumor had the location of the diamond fields along the San Juan near the mouth of Chinle Creek. Although small garnets can be found in the nearby Mule Ear Diatreme, there are no precious gems. Before the end of 1872, the hoax was revealed.

Early in 1879, James Merrick and Ernest Mitchell went searching for an old Indian silver mine supposedly located near the San Juan. They never returned. Their bodies were eventually discovered by the massive red sandstone buttes that now bear their names in Monument Valley. It was believed they had been killed by Indians after locating the mine. Others searched for the silver but to no avail—the legend lives on.

January 5, 1893, the Salt Lake Tribune reported, "7000 men are in the mines. . .600 arrive daily." Gold along the San Juan? Nineteen days later, "No gold has been taken from the river." Some extremely fine, flour gold had been located but most of the men left the area empty-handed. One disappointed prospector inscribed his feelings on a sandstone boulder: "One hundred dollars reward for the damned fool who started the gold boom."

K.W. Trimble   Otis Marston Collection

Besides small quantities of gold, oil had also been discovered in the canyons. In the spring of 1882, E. L. Goodridge, a gold prospector, floated down the San Juan in a homemade wooden skiff and noticed oil seeps near Slickhorn Gulch. Maybe a fortune could be made drilling for oil; but how to get a drilling rig to this isolated spot?

Goodridge built a primitive road into Slickhorn Gulch. He then hauled a drilling engine more than 175 miles from Gallup, New Mexico, taking more than 30 days to reach the canyon. The engine was within yards of the proposed drilling site when it tumbled off the road, over the cliffs, and was broken beyond repair. Goodridge abandoned this drilling site but later in 1908, he did manage to bring in a gusher at Mexican Hat. Unfortunately his well was not considered to be of commercial value at that time.

Other prospectors boating on the San Juan during the late 1890s included Bert Loper, later to become famous as a skilled whitewater boatman, and Walter Mendenhall, who built a cabin in the saddle at Mendenhall Loop.

The Trimble Expedition which mapped the river in 1921 marked the first scientific examination of the canyons of the San Juan. These explorers were studying the geology and potential for hydroelectric dams on the river. Bert Loper was hired as the chief boatman.

"In shooting the worst ones (rapids) Loper, the boatman, was the only member of the party to stay in the boats; the other men walked along the banks around such rapids." H. D. Miser, geologist for the expedition, went on to say, "The voyage was attended by strenuous labor and hardships, such as may always be expected in exploring an unknown canyon with its rapids in an uninhabited region."

Through the years other studies have been made of the river and its resources. During the 1930s, the University of California, the National Park Service, and the Museum of Northern Arizona sponsored the Rainbow Bridge–Monument Valley Expeditions that mapped geology, collected plants and animals, and recorded prehistoric sites in the San Juan area. Before the waters of Lake Powell inundated the lower reaches of the San Juan River in the late 1960s, the Museum of Northern Arizona and the University of Utah conducted emergency archaeological, historical, and biological studies to salvage data soon to be lost under the lake.

One resource no less valuable than gold and no less intriguing than ancient earth history is the beauty of the river and its canyons. In 1934, Norman Nevills built his first boat out of boards from an old privy and floated down the San Juan to the Colorado. Recreational boating on the San Juan was launched.

Over the next 15 years, Nevills guided hundreds of wide-eyed tourists down the San Juan and Colorado rivers. No doubt he would still be manning the oars today if he had not met an untimely death in 1949.

Norman Nevills — Otis Marston Collection

Today 11 outfitters operate a variety of float trips. Additionally, private trips are growing in popularity and the National Park Service and Bureau of Land Management are attempting to insure that the river and its environs are not degraded by overuse.

Other threats to the San Juan include dams, uranium mining, oil drilling, and competition for the river's water for irrigation. The only hope of retaining the wilderness aspects of the San Juan and its canyons is to be vocal and politically active in the decision-making processes affecting the few remaining desert rivers.

FLORA and FAUNA

The tracks of coyotes were numerous in all parts of the canyon and the track of a single mountain lion was observed. The birds noted include black eagles, owls, ducks, blue cranes, swallows, and rock wrens. . . . Several kinds of insects, especially flies, mosquitoes, ants, and scorpions proved to be far too numerous for comfort. . . . Lizards sometimes got into the beds and always caused considerable anxiety for fear they were scorpions, until a match was lighted and thus revealed their harmless nature.

H.D. Miser, 1924

The first-time visitor to the San Juan River country is usually struck by the stark, endless vistas of red and buff sandstone and blue sky. The landscape seems empty, devoid of any living creature, an inhospitable region. It is indeed a place of terrible extremes. Less than 6 inches of rain fall per year yet intricate canyon systems funnel the scant moisture into devastating flash floods. Searing summer heat, when temperatures climb to over 100°F in the shade, contrasts with the winter's arctic cold with bone-chilling nights of −20°F and nothing to break the wind. Moisture that has seeped into cracks within the rocks expands and contracts—the boulders crumble. You have to be tough to endure.

Availability of water is the key to survival in this arid land. A desert dweller must be able to locate water, conserve it efficiently, and have some means to persist through drought periods. Coping with the temperature extremes is also important.

The river's water offers a ribbon of life through the San Juan country and softens some of the harshness of this desert. Examine closely the apparently lifeless land. There are survivors here; yes, even plants and animals that revel in the xeric way of life.

There are four general habitats within the San Juan River canyons; namely, the river itself, the riparian or streamside environment, the river terrace/talus community, and the cliff environment. Each habitat has fairly specific groups of plants that can grow there and to a certain extent particular assemblages of animals. It must be remembered that animals are mobile and range about in their quest for food, shelter, and a mate; they are therefore not necessarily restricted to one particular plant community.

The River

The turbid waters of the San Juan appear to be an unlikely place to harbor fish; yet, about 17 species are believed to exist between Navajo Dam in New Mexico and Clay Hills Crossing on Lake Powell. Of these 17, 6 species are native and especially adapted to living and reproducing

Painting by K. Pendleton,courtesy Colorado Division of Wildlife

Razorback Sucker

in relatively warm, silty water. As the river is dammed, diverted, polluted, or otherwise violated by man, the native fish are disappearing. The razorback sucker has been federally designated threatened and the Colorado River squawfish is classified endangered.

Of the introduced species of fish, the channel catfish and black bullhead catfish are of interest to fishermen. Few people fish these waters; perhaps the cats taste too muddy. The Navajo who live along the San Juan traditionally have a taboo against consuming fish.

Sometimes seen swimming on the river are a variety of ducks and other waterfowl. Common species include green-winged teal, cinnamon teal, western grebe, and Canadian geese. Often skimming over the river's surface are swallows and swifts taking a drink or hawking insects.

The Riparian Community

Along the riverbank is a band of verdant vegetation that is lush in comparison to the sun-baked terraces farther back from the river's influence. This riparian community is typically composed of tamarisk (also called salt cedar, an introduced shrub native to the Mediterranean area), coyote willow, arrowweed, and seep willow. Where the shoreline is marshy, cattails, common reed, and horsetail rush appear. Where the riverbanks are relatively wide, such as between Montezuma Creek and Comb Wash, small riparian woodlands form of Fremont cottonwood, box elder, and the naturalized exotic Russian olive.

In the spring and early summer, the red-spotted and Woodhouse's toads are sometimes seen hopping down the

Painting by K. Pendleton, courtesy Colorado Division of Wildlife

Colorado River Squawfish

Riparian Community

Barbara Kemp

shore to the water's edge. Both are a dull sandy color which blends in well with the beach sand. In the evenings, especially in the side canyons, you may hear the loud "baa-baa" call of the lovesick canyon tree frog. While crashing through the reeds and cattails, you may scare up a western wandering garter snake.

Probably the mammal most in evidence in the riparian zone is the beaver. Many of the cottonwoods and willows and occasionally tamarisk show gnaw marks left by these primarily nocturnal animals. The San Juan beavers do not attempt to build dams on this unpredictable river but rather build their dens in the riverbanks.

Barbara Kemp

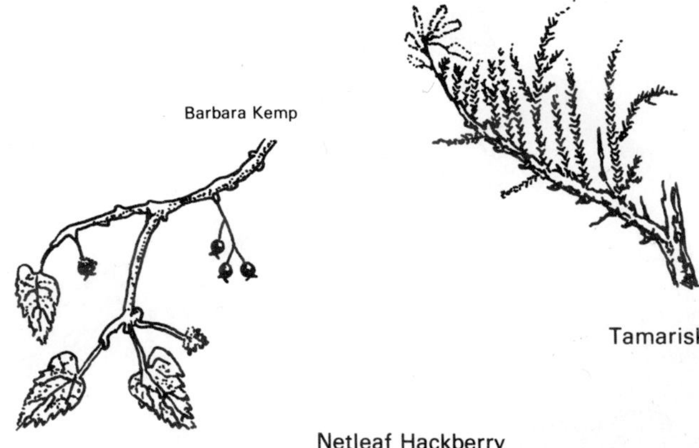

Tamarisk

Netleaf Hackberry

Barbara Kemp

Coyote Willow

Between Montezuma Creek and Comb Wash, it is not unusual to observe great blue herons and snowy egrets wading near shore, stalking minnows and frogs. Running along the beaches are spotted sandpipers and killdeer searching for insects and other invertebrates.

In the riparian shrubs and woodlands occur sparrow hawks (kestrels), mourning doves, ash-throated flycatchers, western kingbirds, red-shafted flickers, Bullock's orioles, Lucy's warblers, yellow warblers, Bell's vireo, house finches, rufous-sided towhees, and the recently introduced ring-necked pheasant.

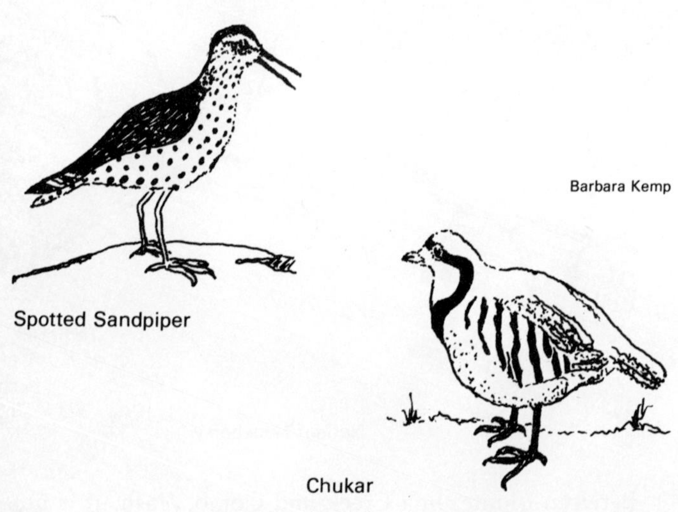

Spotted Sandpiper

Chukar

Terrace/Talus Community

Between the band of riparian vegetation and the vertical cliffs is the terrace or bench and talus communities. The soil varies from silt and sand to clays mixed with rocks. Water is at a premium; the plants must rely almost solely on the scant rainfall. Competition for the precious fluid results in the individual plants being fairly well spaced. Leaves tend to be small and waxy to slow down evaporation, or the leaves may even fall off during drought periods. Root systems are either deep in an attempt to tap secret water reserves or shallow and wide to garner any falling precipitation.

Shrubs in this community include snakeweed, greasewood, four-wing saltbush, barberry, Mormon tea, sagebrush, buffaloberry, blackbrush, rabbitbrush, squawbush, scrub oak and wavy-leaf oak. Banana yucca, narrow-leaf yucca, and the similar looking but taxonomically different Utah agave dot the talus slopes. The most common cactus is the prickly pear but some small cholla and hedgehog cactus can be found. Winters are apparently too cold for more succulent types of cactus to exist here; they would freeze, crack open, and die.

Barbara Kemp

Four-Wing Saltbush

Terrace/Talus Community

Barbara Kemp

The few small trees growing in this community are the Utah and one-seed junipers (incorrectly called cedars by many local cowboys), netleaf hackberry, pinyon pine, Gambel's oak, and single-leaf ash.

When the rains come, usually as winter showers and late summer thunderstorms, quite a few wildflowers and grasses make their debut. Some are Indian rice grass, wild buckwheat, globemallow, groundsel, paintbrush, wild onion, dock or wild rhubarb, evening primrose, sego-lily, trailing four-o'clock, princes plume, and scarlet gilia.

Lizards are by far the most obvious animals in this community. Bright blue-green collared lizards look like miniature dinosaurs and sometimes stand up and run on their hind legs to escape. The slinky snakelike whiptail lizard darts from bush to bush. Around anthills horned lizards (incorrectly called horny toads) may be seen feeding. Sunning on boulders may be the well-armored desert spiny lizard, side-blotched lizard, tree lizard, or fence lizard.

Snakes are rarely encountered. With no legs to raise their tender body off the hot ground, they must wait until dusk to venture out. Common species include the gopher snake, striped whipsnake, and western rattlesnake.

Most of the mammals within this habitat are nocturnal but you may see the grey-colored rock squirrel or the chipmunklike antelope ground squirrel scooting along with its white tail held up over its back.

In sandy areas the burrows of kangaroo rats or pocket mice may be noticed. These two animals avoid the heat of day by staying in their cool tunnels. Both of these rodents feed on seeds that are high in carbohydrates. Water is one of the products of carbohydrate digestion. (This is called metabolic water.) Both possess remarkably efficient kidneys so that little water is lost through urination. Neither emerge from their burrows until the cool and higher relative humidity of evening. Thus they are able to go without drinking free water or eating succulent plant material; their metabolic water fills their needs. Truly an amazing example of adaptation to a desert environment.

Some of the desert animals avoid drought periods by going into a deep sleep trance called estivation. Their metabolic processes slow down to a near deathlike state. Rain and/or cooler temperatures arouse them.

Other residents of the terrace habitat include the canyon mouse, jackrabbit, skunk, ringtail cat, grey fox, coyote, mule deer, and the rare mountain lion. About the only birds that could be considered typical of this community are the introduced quail called chukars, horned larks, black-throated sparrows, sage sparrow, and an occasional Say's phoebe. Lack of tree-nesting sites probably accounts for the paucity of bird life.

Cliff Community

The cliffs are, of course, a dominant topographic feature of the canyon. Many of the same plants found in the terrace community ascend the vertical walls, rooting in cracks or little pockets of soil. Runoff after storms drips down the walls and may encourage datura, birchleaf buckthorn, or poison ivy to grow along the base of a cliff.

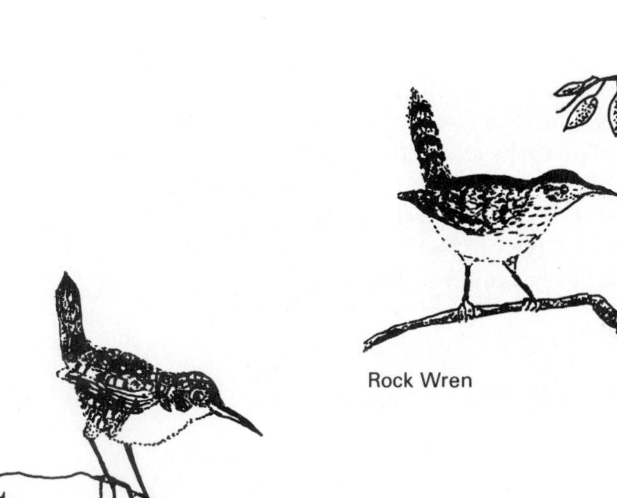

Rock Wren

Canyon Wren

Cliff Community

Barbara Kemp

Barbara Kemp

Cliff Swallow and Nest

Violet-green Swallow

The small lizards, such as the side-blotched, fence, and tree, scurry up and down the walls. Quite a number of bats roost in the small caves and crevices. Wood rats build their middens under the protective overhangs. Rock squirrels and other rodents nest on the rocky cliffs. The master cliff climber, though, is the bighorn sheep. Look for sheep between Mexican Hat and Clay Hills Crossing.

Cliff nesting birds include the cliff swallow whose mud nest looks like an Anasazi wattle-and-daub structure, violet-green swallow, white-throated swift, canyon wren, rock wren, raven, turkey vulture, red-tailed hawk, golden eagle, prairie falcon, and the great horned owl.

The rocks and cliffs provide shelter; the rivers and rare storms give water; from the sun comes energy. A marvelously adapted yet fragile plant and animal community has evolved to meet the demands of a challenging, harsh, and very beautiful desert land.

List of Common Plant and Animal Names Used and Scientific Equivalents

Plants:

Common name	Scientific name
Arrowweed	*Pluchea sericea*
Banana yucca	*Yucca baccata*
Barberry	*Berberis fremontii*
Birchleaf buckthorn	*Rhamnus betulaefolia*
Blackbrush	*Coleogyne ramosissima*
Box elder	*Acer negundo*
Buffaloberry	*Shepherdia rotundifolia*
Cattails	*Typha* spp.
Cholla	*Opuntia* spp.
Common reed	*Phragmites communis*
Coyote willow	*Salix exigua*
Datura	*Datura meteloides*
Dock	*Rumex hymenosepalus*
Evening primrose	*Oenothera* spp.
Four-wing saltbush	*Atriplex canescens*
Fremont cottonwood	*Populus fremontii*
Gambel's oak	*Quercus gambellii*
Globemallow	*Sphaeralcea* spp.
Greasewood	*Sarcobatus vermiculatus*
Groundsel	*Senecio* spp.
Hedgehog cactus	*Echinocereus* spp.
Horsetail rush	*Equisetum* spp.
Indian rice grass	*Oryzopsis* spp.
Mormon tea	*Ephedra* spp.
Narrow-leaf yucca	*Yucca angustissima*
Netleaf hackberry	*Celtis reticulata*
One-seed juniper	*Juniperus monosperma*
Paintbrush	*Castilleja* spp.
Pinyon pine	*Pinus edulis*
Poison ivy	*Rhus radicans*
Prickly pear	*Opuntia* spp.
Princes plume	*Stanleya pinnata*
Rabbitbrush	*Chrysothamnus nauseosus*
Russian olive	*Elaeagnus angustifolia*
Sagebrush	*Artemisia* spp.

Salt cedar	*Tamarix chinensis*
Scarlet gilia	*Gilia aggregata*
Scrub oak	*Quercus turbinella*
Seep willow	*Baccharis* spp.
Sego-lily	*Calochortus nuttallii*
Single-leaf ash	*Fraxinus anomala*
Snakeweed	*Gutierrezia sarothrae*
Squawbush	*Rhus trilobata*
Tamarisk	*Tamarix chinensis*
Trailing four-o'clock	*Allionia incarnata*
Utah agave	*Agave utahensis*
Utah juniper	*Juniperus osteosperma*
Wavy-leaf oak	*Quercus undulata*
Wild buckwheat	*Eriogonum* spp.
Wild onion	*Allium* spp.
Wild rhubarb	*Rumex hymenosepalus*

Animals:

Antelope ground squirrel	*Ammospermophilus leucurus*
Ash-throated flycatcher	*Myiarchus cinerascens*
Bat	Many species
Beaver	*Castor canadensis*
Bell's vireo	*Vireo bellii*
Bighorn sheep	*Ovis canadensis*
Black bullhead catfish	*Ictalurus melas*
Black-throated sparrow	*Amphispiza bilineata*
Bullock's oriole	*Icterus bullockii*
Canada goose	*Branta canadensis*
Canyon mouse	*Peromyscus crinitus*
Canyon tree frog	*Hyla areniclor*
Canyon wren	*Catherpes mexicanus*
Channel catfish	*Ictalurus punctatus*
Chukar	*Alectoris graeca*
Cinnamon teal	*Anas cyanoptera*
Cliff swallow	*Petrochelidon pyrrhonota*
Collared lizard	*Crotaphytus collaris*

Common name	Scientific name
Colorado River squawfish	*Ptychocheilus lucius*
Coyote	*Canis latrans*
Desert spiny lizard	*Sceloporus magister*
Fence lizard	*Sceloporus undulatus*
Golden eagle	*Aquila chrysaetos*
Gopher snake	*Pituophis melanoleucus*
Great blue heron	*Ardea herodias*
Great horned owl	*Bubo virginianus*
Green-winged teal	*Anas carolinensis*
Grey fox	*Urocyon cinereoargenteus*
Horned lark	*Eremophila alpestris*
Horned lizard	*Phrynosoma douglassi*
House finch	*Carpodacus mexicanus*
Jackrabbit	*Lepus californicus*
Kangaroo rat	*Dipodomys* spp.
Kestrel	*Falco sparverius*
Killdeer	*Charadrius vociferus*
Lucy's warbler	*Vermivora luciae*
Mountain lion	*Felis concolor*
Mourning dove	*Zenaidura macroura*
Mule deer	*Odocoileus hemionus*
Pocket mouse	*Perognathus* spp.
Prairie falcon	*Falco mexicanus*
Raven	*Corvus corax*
Razorback sucker	*Xyrauchen texanus*
Red-shafted flicker	*Colaptes cafer*
Red-spotted toad	*Bufo punctatus*
Red-tailed hawk	*Buteo jamaicensis*
Ring-necked pheasant	*Phasianus colchicus*
Ringtail cat	*Bassariscus astutus*
Rock squirrel	*Citellus variegatus*
Rock wren	*Salpinctes obsoletus*
Rufous-sided towhee	*Pipilo erythrophthalmus*
Sage sparrow	*Amphispiza belli*
Say's phoebe	*Sayornis saya*
Side-blotched lizard	*Uta stansburiana*
Skunk	*Mephitis mephitis*
Snowy egret	*Leucophoyx thula*
Sparrow hawk	*Falco sparverius*
Spotted sandpiper	*Actitis macularia*
Striped whipsnake	*Masticophis taeniatus*
Tree lizard	*Urosaurus ornatus*

Turkey vulture	*Cathartes aura*
Violet-green swallow	*Tachycineta thalassina*
Western grebe	*Aechmophorus occidentalis*
Western kingbird	*Tyrannus verticalis*
Western rattlesnake	*Crotalus viridis*
Western wandering garter snake	*Thamnophis elegans*
Whiptail lizard	*Cnemidophorus* spp.
White-throated swift	*Aeronautes saxatalis*
Wood rat	*Neotoma* spp.
Woodhouse's toad	*Bufo woodhousei*
Yellow warbler	*Dendroica petechia*

SUGGESTED READING

Ambler, J. R. 1977, *The Anasazi*. Museum of Northern Arizona, Flagstaff.

Baars, D. L. 1972. *Red Rock Country*. Doubleday Press, Garden City.

Baars, D. L., ed. 1973. *Geology of the Canyons of the San Juan River*. Four Corners Geological Society, Durango.

Baars, D. L. and C. M. Molenaar. 1971. *Geology of Canyonlands and Cataract Canyon*. Four Corners Geological Society, Durango.

Barnes, F. A. 1977. *Canyon Country Hiking and Natural History*. Wasatch Publishers, Inc., Salt Lake City.

Barnes, F. A. and M. Pendleton. 1979. *Canyon Country Prehistoric Indians*. Wasatch Publishers, Inc., Salt Lake City.

Crampton, C. G. 1964. *The San Juan Canyon Historical Sites*. University of Utah Anthropological Papers No. 70, Salt Lake City.

Crampton, C. G. 1973. *Standing Up Country*. Alfred A. Knopf, Inc., New York.

Doolittle, J. 1974. *Canyons and Mesas*. Time-Life Books, New York.

Elmore, F. H. 1976. *Shrubs and Trees of the Southwest Uplands*. Southwest Parks and Monuments Association, Globe.

Gregory, H. E. 1916. *The Navajo Country*. U.S. Geological Survey Professional Paper No. 380, Washington, D.C.

Gregory, H. E. 1938. *The San Juan Country*. U.S. Geological Survey Professional Paper No. 188, Washington, D.C.

Hintze, L. H. 1973. *Geologic History of Utah*. Brigham Young University of Geology Studies, vol. 20, #3, Provo.

Hunt, C. B. 1969. "Geologic History of the Colorado River" *in The Colorado River Region and John Wesley Powell*. U.S. Geological Survey Professional Paper No. 669, Washington, D.C.

Huser, V. 1978. *Canyon Country Paddles*. Wasatch Publishers, Inc., Salt Lake City.

Lee, Katie. 1964. *Folk Songs of the Colorado River*. A phonograph record. Folkways Records, New York.

Lipe, W. D. 1970. "Anasazi Communities on the Red Rock Plateau, Southeastern Utah" *in Reconstructing Prehistoric Pueblo Societies,* edited by W. A. Longacre. School of American Research, Santa Fe and University of New Mexico, Albuquerque.

McNitt, F. 1974. *Richard Wetherill: Anasazi.* University of New Mexico Press, Albuquerque.

Macomb, J. N. 1876. *Report of the Exploring Expedition from Santa Fe, New Mexico, to the junction of the Grand and the Green Rivers of the Great Colorado of the West, in 1859, Under the Command of Capt. J. N. Macomb; with a Geological Report by Prof. J. S. Newberry.* Government Printing Office, Washington, D.C.

Miller, D. E. 1959. *Hole-In-The-Rock.* University of Utah Press, Salt Lake City.

Miser, H. D. 1924. *The San Juan Canyon.* U.S. Geological Survey Water Supply Paper No. 538, Washington, D.C.

Newberry, J. S. 1876. See Macomb, 1876.

Redd, Charles. 1950. "Short Cut to the San Juan" *in 1949 Brand Book.* Denver Posse of the Westerners, Denver.

Rocky Mountain Association of Geologists. 1972. *Geologic Atlas of the Rocky Mountain Region.*

Schaafsma, P. 1971. *The Rock Art of Utah.* Peabody Museum, Cambridge.

Sears, J. D. 1956. *Geology of Comb Ridge and Vicinity North of the San Juan River, San Juan County, UT.* U.S. Geological Survey Bulletin 1021-E, Washington, D.C.

Stegner, W. 1946. *The Sound of Mountain Water.* Doubleday and Company, Inc., Garden City.

Stuart-Alexander, D. E., E. M. Shoemaker, and H. J. Moore. 1972. *Geologic Map of the Mule Ear Diatreme, San Juan County, Utah.* U.S. Geological Survey Misc. Geol. Inves., Map I-674, Washington, D.C.

Zwinger, Ann. 1975. *Run, River, Run.* Harper & Row, Publishers, Inc., New York.

Zwinger, Ann. 1978. *Wind in the Rock.* Harper & Row, Publishers, Inc., New York.

ACKNOWLEDGMENTS

The expertise of many individuals went into the making of this river guide. For providing photographs I would like to thank Michael Collier, David Maren, John Running, and especially the late Otis "Dock" Marston. I also appreciate the help received from Earl Hatfield, Barry Lutz, Stanley Stillion, and the U.S. Geological Survey concerning technical aspects of the river map and photos. Thanks, too, to all those who gave of their time, materials, and special knowledge about the San Juan including Don Baars, Bill Belknap, George Billingsley, C. Gregory Crampton, Dorothy House, Don Keller, Katie Lee, the San Juan Resource Area Office of the Bureau of Land Management, and the National Park Service. The wonderful drawings were done by Barbara Kemp. And finally, to Peter Winn, who taught me about the secrets of the water, and to my wife, Ann, and my good friend Bill Williams, both of whom have shared the adventure, thank you.

SA

CONTRIBUTORS

Stewart Aitchison—a naturalist and guide for several outdoor adventure/educational outfitters. He has authored numerous scientific and popular articles as well as the books *Oak Creek Canyon and the Red Rock Country of Arizona* and *A Naturalist's Grand Canyon Hiking Guide*.

Peter Winn—a graduate geology student at the University of Utah and longtime commercial river-runner.

Don Keller—an archaeologist with the Museum of Northern Arizona and an avid whitewater boatman.